Ilya Gertsbakh

# Measurement Theory for Engineers

T0224102

Springer

Berlin
Heidelberg
New York
Hong Kong
London
Milan
Paris
Tokyo

Ilya Gertsbakh

# Measurement Theory
# for Engineers

With 25 Figures and 49 Tables

 Springer

Ilya Gertsbakh

Ben Gurion University of the Negev
Department of Mathematics
84105 Beer-Sheva
Israel

ISBN 978-3-642-05509-6

Springer-Verlag Berlin Heidelberg New York

Cataloging-in-Publication Data applied for.
Bibliographic information published by Die Deutsche Bibliothek. Die Deutsche Bibliothek lists
this publication in the Deutsche Nationalbibliografie; detailed bibliographic data is available in
the Internet at <http://dnb.ddb.de>.

Springer-Verlag Berlin Heidelberg New York
a member of BertelsmannSpringer Science+Business Media GmbH

http://www.springer.de

© Springer-Verlag Berlin Heidelberg 2010
Printed in Germany

Cover-design: Medio, Berlin
Printed on acid-free paper        62 / 3020 hu – 5 4 3 2 1 0

To my wife Ada

# Preface

The material in this book was first presented as a one-semester graduate course in Measurement Theory for M.Sc. students of the Industrial Engineering Department of Ben Gurion University in the 2000/2001 academic year.

The book is devoted to various aspects of the statistical analysis of data arising in the process of measurement. We would like to stress that the book is devoted to *general* problems arising in processing measurement data and does not deal with various aspects of special measurement techniques. For example, we do not go into the details of how special physical parameters, say ohmic resistance or temperature, should be measured. We also omit the accuracy analysis of particular measurement devices.

The Introduction (Chapter 1) gives a general and brief description of the measurement process, defines the measurand and describes different kinds of the measurement error.

Chapter 2 is devoted to the point and interval estimation of the population mean and standard deviation (variance). It also discusses the normal and uniform distributions, the two most widely used distributions in measurement. We give an overview of the basic rules for operating with means and variances of sums of random variables. This information is particularly important for combining measurement results obtained from different sources. There is a brief description of graphical tools for analyzing sample data. This chapter also presents the round-off rules for data presentation.

Chapter 3 contains traditional material on comparing means and variances of several populations. These comparisons are typical for various applications, especially in comparing the performance of new and existing technological processes. We stress how the statistical procedures are affected by measurement errors. This chapter also contains a brief description of Shewhart charts and a discussion on the influence of measurement errors on their performance.

When we measure the output parameter of a certain process, there are two main sources of variability: that of the process and that due to the errors introduced by the measurement process. One of the central issues in statistical measurement theory is the estimation of the contribution of each of these two sources to the overall variability. This problem is studied in the framework of ANOVA with random effects in Chapter 4. We consider one-way ANOVA, repeatability and reproducibility studies in the framework of two-way ANOVA and hierarchical design of experiments.

Very often, it is not possible to carry out repeated measurements, for example, in destructive testing. Then a special design of experiment is necessary to process the measurement data. Two models dealing with this situation are considered in Chapter 4: one is a combination of two hierarchical designs, and another is the Grubbs model.

There are two principal types of measurements: direct and indirect. For example, measuring the voltage by a digital voltmeter can be viewed as a direct measurement: the scale reading of the instrument gives the desired result. On the other hand, when we want to measure the specific weight of some material, there is no such device whose reading would give the desired result. Instead, we have to measure the weight $W$ and the volume $V$, and express the specific weight $\rho$ as their ratio: $\rho = W/V$. This is an example of an indirect measurement. The question of principal importance is estimating the *uncertainty* of the measure of specific weight introduced by the uncertainties in measuring $W$ and $V$. The uncertainty in indirect measurements is estimated by using the so-called *error propagation formula*. Its derivation and use are described in Chapter 5.

Chapter 6 is devoted to *calibration* of measurement instruments. In many instances, we are not able to measure directly the parameter of interest, $x$. Instead, we are able to measure some other parameter, say $y$, which is related to $x$ via some unknown relationship $y = \phi(x)$. When we observe the value of $y = y_0$ we must find out the corresponding value of $x$ as $x_0 = \phi^{-1}(y_0)$. The statistician is faced with the problem of estimating the unknown function $\phi(\cdot)$ and the uncertainty in the value of $x_0$. These problems are known as "calibration" and "inverse regression". We present the results for linear calibration curves with equal and variable response variances, and with uncertainties in both variables $x$ and $y$.

It is a well-known fact from measurement practice that if similar samples are analyzed independently by several laboratories, the results might be very diverse. Chapter 7 deals with statistical analysis of data obtained by several laboratories, so-called, "collaborative studies". The purpose of these is to locate laboratories which have large systematic and/or random measurement errors.

Chapter 8 is devoted to the study of two special situations arising in measurements. One is when the measurand variability is of the same magnitude as the measurement instrument scale unit. Then the repeated measurements are either identical or differ only by one scale unit. The second is measurements under constraints on the measured parameters.

All chapters, except Chapter 7, have exercises, most with answers and solutions, to enable the reader to measure his/her progress.

The book assumes that the reader already has some acquaintance with probability and statistics. It would also be highly desirable to have some knowledge of the design of experiments. In particular, I assume that such notions as mean, variance, density function, independent random variables, confidence interval, the *t*-test and ANOVA, are familiar to the reader.

I believe that the book might serve as a text for graduate engineering students interested in measurements, as well as a reference book for researchers and engineers in industry working in the field of quality control and in measurement laboratories.

I would like to express my deep gratitude to Dr. E. Tartakovsky for many valuable discussions on measurement and for introducing me to some aspects of statistical measurement theory.

Ilya Gertsbakh
Beersheva, October 2002

# Contents

# Chapter 1

# Introduction: Measurand and Measurement Errors

*Truth lies within a little and certain compass, but error is immense.*

Bolingbroke

## 1.1 Measurand and Measurement Errors

Our starting point will be a well-defined physical object which is characterized by one or more properties, each of a quantitative nature.

Consider for example a cylinder steel shaft, a rolled steel sheet, and a specimen made of silver. A complete characterization of any of these objects demands an infinite amount of information. We will be interested, for the sake of simplicity, in a single (i.e. one-dimensional) parameter or quantity. For example we will be interested

**a.** in *the diameter* of the cylinder shaft measured in its midsection;

**b.** in *the thickness* of the rolled steel sheet, or in the chromium proportion in the steel;

**c.** in a physical constant termed the *specific weight*, measured in grams per cubic centimeter of silver.

The quantity whose value we want to evaluate is called the *measurand*. So the diameter of the cylinder shaft in its midsection is the measurand in example **a.** The proportion of chromium in the rolled steel sheet is the measurand for example **b.** The measurand in example **c** is the specific weight of silver.

We define *measurement* as the assignment of a number to the measurand, using special technical means (measuring instruments) and a specified technical procedure. For example, we use a micrometer to read the diameter of the shaft, according to the accepted rules for operating the micrometer. The measurement result is 7.252 mm. Chemical analysis carried out according to specified rules establishes that the rolled steel sheet has 2.52% chromium. Weighting the piece of silver and measuring its volume produces the result 10.502 g/cm$^3$. The word "measurement" is also used to describe the set of operations carried out to determine the measurand.

Note that we restrict our attention in this book to measuring well-defined physical parameters and do not deal with measurements related to psychology and sociology, such as IQ, level of prosperity, or inflation.

Suppose now that we *repeat* the measurement process. For example, we measure the shaft midsection diameter several times; or repeat the chemical analysis for percentage of chromium several times, taking different samples from the metal sheet; or repeat the measurements of weight and volume needed to calculate the specific weight of silver. The fundamental fact is that each repetition of the measurement, as a rule, will produce a different result.

We can say that the measurement results are subject to variations. An equivalent statement is that the result of any measurement is subject to *uncertainty*. In principle, we can distinguish two sources of uncertainty: the variation created by the changes in the measurand and the variations which are intrinsic to the measurement instrument and/or to the measurement process.

As a rule, we ignore the possibility of an interaction between the measurand and the measurement instrument or assume that the effect of this interaction on the measurement result is negligible. For example, in the process of measuring the diameter, we apply a certain pressure which itself may cause a deformation of the object and change its diameter. A good example of such interaction is measuring blood pressure: most people react to the measurement procedure with a raise in blood pressure.

One source of uncertainty (variability) is the measurand itself. Consider for example the cross-section of the cylindric shaft. It is not an ideal circle, but similar, in exaggerated form, to Fig. 1.1. Obviously, measurements in different directions will produce different results, ranging, say from 10.252 mm to 10.257 mm.

We can try to avoid the "multiplicity" of results by redefining the measurand: suppose that the diameter is defined as the average of two measurements taken in perpendicular directions as shown in Fig. 1.1. Then the newly defined diameter will vary from one shaft to another due to the variations in the process of shaft production.

The situation with the rolled steel sheet is similar. Specimens for the purpose of establishing the proportion of chromium are taken from different locations on the sheet, and/or from different sheets. They will contain, depending on the properties of the production process, different amounts of chromium ranging from, say, 2.5% to 2.6%.

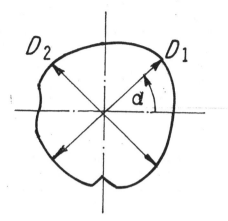

Figure 1.1. The diameter $D$ of the shaft depends on $\alpha$.

The situation with the specific weight of silver is somewhat different. We postulate that the specific weight of silver is not subject to change and is a *physical constant*. (Of course, this physical constant is defined for a specified range of ambient temperatures and possibly other environmental factors, such as altitude.) If we repeat the process of measuring the weight and the volume of the same piece of silver, under prescribed stable environmental conditions, we nevertheless will get different results.

How can this be explained? Let us examine the measurement process. Here we have what are called *indirect* measurements. There is no instrument which can provide the value of specific weight. To obtain the specific weight, we have to measure the weight and the volume of our specimen. The repeated measurements of the weight, even if carried out on the same measurement device, will have small variations, say of magnitude 0.0001 g. Random vibrations, temperature changes, reading errors etc. are responsible for these variations. The repeated measurements of the specimen volume are also subject to variations, which may depend on the measurement method. It may happen that these variations are of magnitude, say, 0.0001 cm$^3$. Thus the repeated measurements of the specific weight, which is the ratio weight/volume, will also vary from measurement to measurement. Note that the measurand itself remains constant, unchanged.

In measuring the diameter of the cylindrical shaft, the measurement results are influenced both by the variations in the "theoretical" diameter, and by inaccuracies caused by the measurement itself.

The relationship between the "true" value of the measurand and the result of the measurement can be expressed in the following form:

$$Y = \mu + \epsilon,$$ 
(1.1.1)

where $\mu$ is the true value of the physical constant and $\epsilon$ is the measurement error.

It should be stressed that (1.1.1) is valid for the situation when there exists a well-defined "true" value of the measurand $\mu$. How to extend (1.1.1) to the situation in which the measurand itself is subject to variations, as it is in our shaft diameter and chromium content examples?

Let us adopt the approach widely used in statistics. Introduce the notion of a *population* as the totality of all possible values of the shaft diameters. (We have in mind a specified shaft production process operating under fixed, controlled conditions.) Fisher (1941) calls this totality a "hypothetical infinite population" (see Mandel (1994), p. 7). The value $D$ of the diameter of each particular shaft therefore becomes a *random variable*. To put it simply, the result of each particular measurement is unpredictable, it is subject to random variations. What we observe as the result of measuring several randomly chosen shafts becomes a *random sample* taken from the above infinite population which we create in our mind.

How, then, do we define the measurand? The measurand, by our definition, is the *mean* value of this population, or more formally, the mean value $\mu$ of the corresponding random variable $D$:

$$\mu = E[D]. \tag{1.1.2}$$

Here $E[\cdot]$ denotes the expectation (the mean value).

Some authors define the measurand as "A value of a physical quantity to be measured" (see Rabinovich 2000, p. 287). Our definition is in fact wider, and it includes also population parameters, e.g. the population mean. On the other hand, we can say that the population mean is considered as a "physical quantity".

The purpose of measuring is not always to assign a single numerical value to the measurand. Sometimes the desired measurement result is an *interval*. Consider for example measuring the content of a toxic element, say lead, in drinking water. The quantity of practical interest is the mean lead content. Since this varies in time and space, in addition to having a point estimate of the mean value, the water specialists would also like to have an interval-type estimate of the mean, e.g. a confidence interval for the mean. We recall that the $1 - 2\alpha$ confidence interval, by definition, contains the unknown parameter of interest with probability $P = 1 - 2\alpha$.

It is always possible to represent the measurement result $Y$ as

$$Y = \mu + X, \tag{1.1.3}$$

where the random variable $X$ reflects all uncertainty of the measurement result.

We can go a step further and try to find out the structure of the random variable $X$. We can say that $X$ consists of two principal parts: one, $X_0$, reflects the variations of the measurand in the population, i.e. its deviations from the mean value $\mu$. Imagine that our measurement instruments are free of any errors, i.e. they are absolutely accurate. Then we would observe in repeated measurements only the realizations of the random variable $\mu + X_0$.

The second principal component of $X$ is the "pure" measurement error $\epsilon$. This is introduced by the measurement instrument and/or by the whole measurement process. Rabinovich (2000, p. 286) defines the error of measurement as "the deviation of the result of a measurement from the true value of the measurand".

Now the measurement result can be represented in the following form:

$$Y = \mu + X_0 + \epsilon. \qquad (1.1.4)$$

What can be said about the random variable $\epsilon$? There is an ample literature devoted to the study of the structure of this random variable in general, and for specific measurement instruments in particular. Rabinovich (2000) assumes that $\epsilon$ consists of elementary errors of different type.

The first type is so-called *absolutely constant* error that remains the same in repeated measurements performed under the same conditions. By their nature these errors are systematic. For example, a temperature meter has a nonlinear characteristic of a thermocouple which is assumed to be linear, and thus a fixed error of say 1°C is present in each temperature measurement in the interval 500–600. So, $\epsilon_1 = 1$.

The second type is *conditionally constant* error. This may be partially systematic and partially random. It remains constant for a fixed instrument, but varies from instrument to instrument. An example is the bias of the zero point of the scale in spring-type weights. A particular instrument may add some weight $\epsilon_2$ to any measurement (weighing) result. The value of the bias changes from instrument to instrument, and certain limits for this error can usually be given in advance.

The third type of error is *random* error. In the course of repeated measurements carried out on *the same* object, by the same operator, under stable conditions, using the same instrument, random errors vary randomly, in an unpredictable way. Operator errors, round-off errors, errors caused by small changes in environmental conditions (e.g. temperature changes, vibrations) are examples of random errors.

Measurement errors themselves are subject to changes due to instrument mechanical wear-out and time drift of other parameters. For example, certain electric measurements are carried out using a built-in source of electricity. A change in the voltage might cause the appearance of an additional error in measurements.

*Remark 1*

For the purpose of measurement data processing, it is often convenient to decompose the measurement error $\epsilon$ into two components $\epsilon_1$ and $\epsilon_2$ which we will call "systematic" and "random", respectively. By definition, the systematic component remains constant for repeated measurements of the same object carried out on the same measurement instrument under the same conditions. The random component produces replicas of independent identically distributed random variables for repeated measurements made by the same instrument under identical conditions.

In theoretical computations, it is often assumed that the first component has a uniform distribution, whose limits can be established from the certificate of the measurement instrument or from other sources. The random component is modeled typically by a zero-mean normal random variable; its variance is either known from experience and/or certificate instrument data, or is estimated from the results of repeated measurements.

Suppose that we carry out a series of repeated measurements of the same measurand $\mu$ on the same instrument, by the same operator and under the same environmental conditions. The result of the $i$th measurement will be

$$Y_i = \mu + \epsilon_1 + \epsilon_{2,i}, \quad i = 1, \ldots, n. \tag{1.1.5}$$

Here $\epsilon_1$ is some unknown quantity which represents the "systematic" error, for example, an error of the measurement instrument. $\epsilon_1$ does not change from measurement to measurement. $\epsilon_{2,i}$, $i = 1, \ldots, n$, are independent random variables and they represent the random errors.

Suppose that we take the average of these $n$ measurements. Then we have

$$\overline{Y} = \mu + \epsilon_1 + \frac{\sum_{i=1}^{n} \epsilon_{2,i}}{n}. \tag{1.1.6}$$

The systematic error is not "filtered out" by the averaging. However, the ratio $a = \sum_{i=1}^{n} \epsilon_{2,i}/n$ tends with probability one to zero as $n$ increases. This fact is known in probability theory as the law of large numbers. An explanation of this phenomenon lies in the fact that the variance of $a$ tends to zero as $n$ goes to infinity.

Equation (1.1.1) is a particular case of (1.1.4). From now on, we use the following formula:

$$Y = \mu + X, \tag{1.1.7}$$

where $\mu$ is the measurand, and the random variable $X$ represents the entire uncertainty of the measurement, i.e. the random and systematic measurement errors, random variations in the population around $\mu$, etc.

*Remark 2*

In the literature, we very often meet the notions of "precision" and "accuracy" (e.g. Miller and Miller 1993, p. 20). The easiest way to explain these notions is to refer to the properties of the random variable $X$ in (1.1.7). Let $\delta$ be the mean of $X$ and $\sigma$ its standard deviation (then $\sigma^2$ is the variance of $\epsilon$. This fact will be denoted as $\epsilon \sim (\delta, \sigma^2)$.) Then measurements with small $\delta$ are called "accurate"; measurements with small $\sigma$ are called "precise". It is very often said that the presence of $\delta \neq 0$ signifies so-called *systematic* error.

Suppose, for example, that we measure the shaft diameter by a micrometer which has a shifted zero point, with a positive shift of 0.020. The measurement results are the following: 10.024, 10.024, 10.025, 10.023, 10.025. This situation reflects precise but nonaccurate measurements.

*Remark 3*
There is a scientific discipline called *metrology* which studies, among other issues, the properties of measurement instruments and measurement processes. One of the central issues in metrology is the analysis of *measurement errors*. Generally speaking, the measurement error is the difference between the "true" value of the measurand (which is assumed to be constant) and the measurement result. Even for a relatively simple measurement instrument, say a digital ohmmeter, it is a quite involved task to find all possible sources of errors in measuring the ohmic resistance and to estimate the limits of the total measurement error. We will not deal with such analysis. The interested reader is referred to Rabinovich (2000).

The limits of maximal error of a measurement instrument (including systematic and random components) usually are stated in the certificate of the instrument, and we may assume in analyzing the measurement data that this information is available.

The data on instrument error (often termed "accuracy") is given in an absolute or relative form. For example, the certificate of a micrometer may state that the maximal measurement error does not exceed 0.001 mm (1 micron). This is an example of the "absolute form". The maximal error of a voltmeter is typically given in a relative form. For example, its certificate may state that the error does not exceed 0.5% of a certain reference value.

# 1.2 Exercises

**1.** Take a ruler and try to measure the area of a rectangular table in $cm^2$ in your room. Carry out the whole experiment 5 times and write down the results. Do they vary? Describe the measurand and the sources of variability of the results. How would you characterize the measurement process in terms of accuracy and precision?

**2.** During your evening walk, measure the distance in steps from your house to the nearest grocery store. Repeat the experiment five times and analyze the results. Why do the results vary?

**3.** Take your weight in the morning, before breakfast, four days in a row. Analyze the results. What is the measurand in this case? Are the measurements biased?

**4** . Consider using a beam balance to weigh an object. An unknown weight $x$ is put on the left balance platform and equalized with a known weight $P_1$. The lengths of the balance arms may not be exactly equal, and thus a systematic error appears in the measurement result. Let $l_1$ and $l_2$ be the unknown lengths of the left and right arm, respectively. Gauss suggested a method of eliminating the systematic error by weighing first on the left platform, and then on the right platform. Suppose that the balancing weights are $P_1$ and $P_2$, respectively. Derive the formula for estimating the weight $x$.

*Hint.* $x \cdot l_1 = P_1 \cdot l_2$. When the weight $x$ is put on the right platform, $x \cdot l_2 = P_2 \cdot l_1$. Derive that $x = \sqrt{P_1 \cdot P_2}$.

# Chapter 2

# Measuring Population Mean and Standard Deviation

*There is no such thing as chance. We have invented this word to express the known effect of every unknown cause.*

Voltaire

## 2.1 Estimation of Mean and Variance

*Example 2.1.1: The weight of Yarok apples*

Apples of a popular brand called "Yarok" are automatically sorted by a special device which controls their size and eliminates those with surface defects. The population in our case is the totality of weight measurements of the daily batch of apples sorted by the device. The experiment consists of weighing a random sample of 20 apples taken from the batch. The results are summarized in Table 2.1.

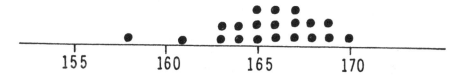

Figure 2.1. Representation of apple weights in form of a dot diagram

Table 2.1: The weights of 20 apples

| $i$ | Weight in grams | $i$ | Weight in grams |
|----|----|----|----|
| 1 | 168 | 11 | 164 |
| 2 | 169 | 12 | 162 |
| 3 | 165 | 13 | 164 |
| 4 | 167 | 14 | 166 |
| 5 | 158 | 15 | 167 |
| 6 | 166 | 16 | 168 |
| 7 | 163 | 17 | 169 |
| 8 | 166 | 18 | 165 |
| 9 | 165 | 19 | 170 |
| 10 | 163 | 20 | 167 |

It is a golden rule of statistical analysis first of all to try to "see" the data. Simple graphs help to accomplish this task. First, we represent the data in the form of a so-called dot-diagram; see Fig. 2.1.

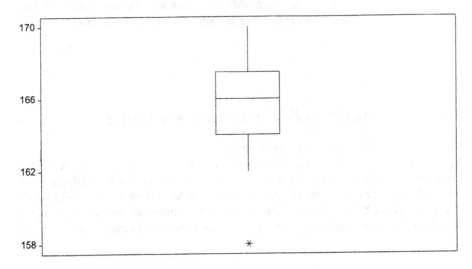

Figure 2.2. Box and whisker plot for the data of Example 2.1.1

A second very popular and useful type of graph is so-called box and whisker plot. This is composed of a box and two whiskers. The box encloses the middle half of the data. It is bisected by a line at the value of the median. The vertical lines at the top and the bottom of the box are called the whiskers, and they indicate the range of "typical" data values. Whiskers always end at the   value of an actual data point and cannot be longer than 1.5 times the size of the box. Extreme values are displayed as "⋆" for possible outliers and "O" for probable outliers. Possible outliers are values that are outside the box boundaries by more than 1.5 time the size of the box. Probable outliers are values that are outside the box boundaries by more than 3 times the size of the box.

In our example, there is one possible outlier – 158 gram. In Sect. 2.5, we describe a formal procedure for identifying a single outlier in a sample.

The mathematical model corresponding to the above example is the following:

$$Y = \mu + X, \tag{2.1.1}$$

where $\mu$ is the mean value for the weight of the population of Yarok apples (the measurand), and the random variable $X$ represents random fluctuation of the weight around its mean. $X$ also incorporates the measurement error.

$X$ is a random variable with mean $\delta$ and variance $\sigma^2$. We use the notation

$$X \sim (\delta, \sigma^2). \tag{2.1.2}$$

Consequently, $Y$ has mean $\mu + \delta$ and variance $\sigma^2$:

$$Y \sim (\mu + \delta, \sigma^2). \tag{2.1.3}$$

Expression (2.1.3) shows that the measurement results include a constant error $\delta$, which might be for example the result of a systematic bias in the apple weighing device. In principle, we cannot "filter out" this systematic error by statistical means alone. We proceed further by assuming that the weighing mechanism has been carefully checked and the bias has been eliminated, i.e. we assume that $\delta = 0$.

*Remark 1*

In chemical measurements, the following method of *subtraction* is used to eliminate the systematic bias in weighing operations. Suppose we suspect that the weighing device has a systematic bias. Put the object of interest on a special pad and measure the weight of the object together with the pad. Then weigh the pad separately, and subtract its weight from the previous result. In this way, the systematic error which is assumed to be the same in both readings will be canceled out.

Recall that $\sigma$, the square root of the variance, is called the *standard deviation*. $\sigma$ is the most popular and important measure of uncertainty which characterizes the spread around the mean value in the population of measurement results. Our purpose is to estimate the mean weight $\mu$ (the measurand) and $\sigma$, the measure of uncertainty. It is important to mention here that $\sigma$ (unlike $\sigma^2$) is measured in the same units as $\mu$. So, in our example $\sigma$ is expressed in grams.

We use capital italic letters $X, Y, Z, \ldots$ to denote random variables and the corresponding small italic letters $x, y, z, \ldots$ to denote the observed values of these random variables.

Let $y_i$ be the observed values of apple weights, $i = 1, \ldots, 20$. In measurements, the unknown mean $\mu$ is usually estimated by *sample mean* $\bar{y}$:

$$\bar{y} = \frac{\sum_{i=1}^{20} y_i}{20}. \qquad (2.1.4)$$

(Often, the word *sample average* or simply *average* is used for the sample mean.) The standard deviation $\sigma$ is estimated by the following quantity $s$:

$$s = \sqrt{\frac{\sum_{i=1}^{20} (y_i - \bar{y})^2}{19}}. \qquad (2.1.5)$$

An estimate of a parameter is often denoted by the same letter with a "hat". So, $\hat{\mu}$ is used to denote an estimate of $\mu$.

For our example we obtain the following results:

$$\bar{y} = \hat{\mu} = 165.6;$$

$$s = 2.82.$$

In general, for a sample $\{y_1, y_2, \ldots, y_n\}$ the estimate of the mean (i.e. the sample mean) is defined as

$$\bar{y} = \frac{\sum_{i=1}^{n} y_i}{n}, \qquad (2.1.6)$$

and the estimate of the standard deviation is defined as

$$s = \sqrt{\frac{\sum_{i=1}^{n} (y_i - \bar{y})^2}{n - 1}}. \qquad (2.1.7)$$

The sample mean is an unbiased (if $\delta = 0$) and consistent estimate of the parameter $\mu$. Recall that the term "consistency" means that the estimate has a variance and bias which both tend to zero as the sample size $n$ goes to infinity.

The sample variance is computed by the formula

$$s^2 = \frac{\sum_{i=1}^{n} (y_i - \bar{y})^2}{n - 1}. \qquad (2.1.8)$$

It is an unbiased and consistent estimate of $\sigma^2$. Using formula (2.1.7) for estimating $\sigma$ introduces a bias, which tends to zero as the sample size goes to infinity.

Sometimes we have a sample of only two observations, $\{y_1, y_2\}$. In this case, (2.1.8) takes the form

$$s^2 = \frac{(y_1 - y_2)^2}{2}. \qquad (2.1.9)$$

*Remark 2: Sheppard's correction*
The result of weighing apples was given as a discrete integer variable, while the weight is a continuous variable. Suppose that the weight reading is provided by a digital device which automatically gives the weight in grams as an integer. The weighing device perceives the "true" weight $y$ and rounds it off to the nearest integer value. What the device does can formally be described as adding to the weight $y$ some quantity $\gamma$, such that $y + \gamma$ is an integer. If the scale unit is $h$ (in our case $h$ is 1 gram), then $\gamma$ lies in the interval $[-h/2, h/2]$. A quite reasonable assumption is that $\gamma$ is *uniformly distributed* in the interval $[-h/2, h/2]$. We use the following notation: $\gamma \sim U(-h/2, h/2)$. We will say more about the uniform distribution in Sect. 2.4. Let us recall that the variance of $\gamma$ is

$$\mathrm{Var}[\gamma] = \frac{h^2}{12}.$$
(2.1.10)

and that the mean value of $\gamma$ is zero: $E[\gamma] = 0$.

Suppose that we observe a sample from the population of $Y = X + \gamma$. $\mu_X$ and $\mu_Y$ are the mean values of $X$ and $Y$, respectively. Since $E[\gamma] = 0$, $\mu_X = \mu_Y = \mu$. The sample mean of random variable $Y$, $\bar{y}$, is therefore an unbiased estimate of $\mu$.

How to estimate from the sample of the observed $Y$ values the variance of the "unobserved" random variable $X$? Under quite wide assumptions regarding the distribution of $X$, it can be shown that

$$\mathrm{Var}[Y] \approx \mathrm{Var}[X] + h^2/12,$$
(2.1.11)

and therefore

$$\mathrm{Var}[X] \approx \mathrm{Var}[Y] - h^2/12.$$
(2.1.12)

Thus, an estimate of $\mathrm{Var}[X]$ can be obtained from the formula

$$s_X^2 = \frac{\sum_{i=1}^{n}(y_i - \bar{y})^2}{n-1} - \frac{h^2}{12}.$$
(2.1.13)

The term $h^2/12$ in this formula is called the Sheppard's correction; see Cramér (1946, Sect. 27.9). The applicability of (2.1.13) rests on the assumption that $\gamma$ has a uniform distribution and the density function for $X$ is "smooth".

Let us compute this correction for our apple example. By (2.1.5), $s_Y^2 = 2.82^2 = 7.95$. $h^2/12 = 1/12 = 0.083$. By (2.1.13) we obtain $s_X^2 = 7.95 - 0.083 = 7.867$, and $s_X = 2.80$. We see that in our example the Sheppard correction makes a negligible contribution.

## Properties of Sample Mean, Sample Variance and the Variance of a Sum of Random Variables: a Reminder

Suppose that the weighing experiment is repeated $m$ times in similar conditions, i.e. we take $m$ random samples of 20 apples and for each of them compute the values of $\overline{y}$ and $s$. For each sample we will obtain *different* results, reflecting the typical situation in *random* sampling.

The apple weights $y_1, \ldots, y_{20}$ are the observed values of random variables $Y_1, Y_2, \ldots, Y_{20}$. These random variables are identically distributed and therefore have the same mean value $\mu$ and the same standard deviation $\sigma$. Moreover, we assume that these random variables are *independent*. The observed sample means $\overline{y}^{(j)}$, $j = 1, 2, \ldots, m$, are the observed values of the random variable

$$\overline{Y} = \frac{\sum_{i=1}^{20} Y_i}{20}. \tag{2.1.14}$$

This random variable is called an *estimator* of $\mu$.

Of great importance is the relationship between $\overline{Y}$ and $\mu$. The mean value of $\overline{Y}$ equals $\mu$:

$$E[\overline{Y}] = \mu, \tag{2.1.15}$$

which means that "on average", the values of the estimator coincide with $\mu$. In statistical terminology, the sample mean $\overline{Y}$ is an *unbiased* estimator of $\mu$. The observed (i.e. the sample) value of the estimator is called *estimate*.

How close will the observed value of $\overline{Y}$ be to $\mu$? This is a question of fundamental importance. The simplest answer can be derived from the following formula:

$$\text{Var}[\overline{Y}] = \frac{\sigma^2}{20}, \tag{2.1.16}$$

i.e. the variance of $\overline{Y}$ is inverse proportional to the sample size $n = 20$. As the sample size $n$ increases, the variance of $\overline{Y}$ decreases and the estimator becomes closer to the unknown value $\mu$. This property defines a so-called *consistent* estimator. (More formally, as $n \to \infty$, then $\overline{Y}$ tends to $\mu$ with probability 1).

Let $n$ be the sample size. It can be proved that the quantity

$$S^2 = \frac{\sum_{i=1}^{n} (Y_i - \overline{Y})^2}{n-1} \tag{2.1.17}$$

is an *unbiased* estimator of the variance $\sigma^2$, i.e. $E[S^2] = \sigma^2$.

We have already mentioned that the square root of $S^2$,

$$S = \sqrt{\frac{\sum_{i=1}^{n} (Y_i - \overline{Y})^2}{n-1}}, \tag{2.1.18}$$

is used as an estimator of the standard deviation $\sigma$. Unfortunately, $S$ is *not* an unbiased estimator of $\sigma$, but in measurement practice we usually use this estimator. As $n$ grows, the bias of $S$ becomes negligible.

If the distribution of $Y_i$ is known, it becomes possible to evaluate the bias of the estimator (2.1.18) and to eliminate it by means of a correction factor. In Sect. 2.6 we consider this issue for the normal distribution.

Readers of the statistical literature are often confused by the similarity of the notation for random variables (estimators) and their observed sample values (estimates). Typically, we use capital letters, like $\overline{Y}$, $S^2$ and $S$ for random variables and/or estimators, as in (2.1.14), (2.1.17) and (2.1.18). When the random variables $Y_i$ in these expressions are replaced by their observed (i.e. sample) values, the corresponding *observed* values of $\overline{Y}$, $S^2$ and $S$ (i.e. their estimates) will be denoted by lower-case letters $\bar{y}$, $s^2$, $s$, as in (2.1.4), (2.1.8) and (2.1.7).

A very important subject is computation of the mean and variance of a sum of random variables. Let $Y_i \sim (\mu_i, \sigma_i^2)$, $i = 1, 2, \ldots, k$, be random variables. Define a new random variable $W = \sum_{i=1}^{k} a_i Y_i$, where $a_i$ are any numbers. It is easily proved that

$$E[W] = \sum_{i=1}^{k} a_i \mu_i. \tag{2.1.19}$$

The situation for the variance is more complex:

$$\text{Var}[W] = \sum_{i=1}^{k} a_i^2 \sigma_i^2 + 2 \sum_{i<j} \text{Cov}[Y_i, Y_j] a_i a_j, \tag{2.1.20}$$

where $\text{Cov}[\cdot]$ is the covariance of $Y_i$ and $Y_j$,

$$\text{Cov}[Y_i, Y_j] = E[(Y_i - \mu_i)(Y_j - \mu_j)]. \tag{2.1.21}$$

It is important to note that if $Y_i$ and $Y_j$ are independent, then their covariance is zero. In most applications we will deal with the case of independent random variables $Y_1, \ldots, Y_k$. Then (2.1.20) becomes

$$\text{Var}[W] = \sum_{i=1}^{k} a_i^2 \sigma_i^2. \tag{2.1.22}$$

This formula has important implications. Suppose that all $Y_i \sim (\mu, \sigma^2)$. Set $a_i = 1/k$ in (2.1.22). Then we obtain that

$$\text{Var}[\frac{\sum_{i=1}^{k} Y_i}{k}] = \frac{\sigma^2}{k}. \tag{2.1.23}$$

Formula (2.1.16) is a particular case of (2.1.23) for $k = 20$.

Now consider the case that all $Y_i$ have the same mean $\mu$ but different variances: $Y_i \sim (\mu, \sigma_i^2)$. (The $Y_i$ are assumed to be independent.) Let us compute the mean and the variance of the random variable

$$W = a_1 Y_1 + a_2 Y_2 + \ldots + a_k Y_k, \tag{2.1.24}$$

where all $a_i$ are nonnegative and sum to 1:

$$\sum_{i=1}^{k} a_1 = 1, \ a_i \geq 0. \tag{2.1.25}$$

Obviously, from (2.1.22) it follows that

$$E[W] = \mu, \quad \text{Var}[W] = \sum_{i=1}^{k} a_i^2 \sigma_i^2. \tag{2.1.26}$$

It follows from (2.1.26) that $W$ remains an unbiased estimator of $\mu$.

Suppose that the choice of the "weights" $a_i$ (under the constraint $\sum a_i = 1$) is up to us. How should $a_i$ be chosen to minimize the variance of $W$? The answer is given by the following formula (we omit its derivation):

$$a_i^\star = \frac{1}{\sigma_i^2 \sum_{j=1}^{k} 1/\sigma_j^2}, \quad i = 1, \dots, k. \tag{2.1.27}$$

The following example demonstrates the use of this formula in measurements.

*Example 2.1.2: Optimal combination of measurements from different instruments*

The same unknown voltage was measured by three different voltmeters A,B and C. The results obtained were $v_A = 9.55$ V; $v_B = 9.50$ V and $v_C = 9.70$ V. The voltmeter certificates claim that the measurement error does not exceed the value of $\Delta_A = 0.075$ V, $\Delta_B = 0.15$ V and $\Delta_C = 0.35$ V, for A, B and C, respectively.

We assume that the measurement error $V_0$ for a voltmeter has a uniform distribution on $(-\Delta, \Delta)$. Then the variance of the corresponding measurement error is equal to

$$\text{Var}[V_0] = \frac{(2\Delta)^2}{12} = \frac{\Delta^2}{3}. \tag{2.1.28}$$

Denote by $V_A, V_B, V_C$ the random measurement results of the voltmeters. The corresponding variances are: $\text{Var}[V_A] = 1.88 \cdot 10^{-3}, \text{Var}[V_B] = 0.0075$ and $\text{Var}[V_C] = 0.041$.

According to (2.1.27), the optimal weights are: $a_A^\star = 0.77$, $a_B^\star = 0.19$, $a_C^\star = 0.04$. Therefore, the minimal-variance unbiased estimate of the unknown voltage is

$$v^\star = 0.77 \cdot 9.55 + 0.19 \cdot 9.5 + 0.04 \cdot 9.7 = 9.546.$$

The corresponding variance is

$$\text{Var}[V^\star] = 0.77^2 \cdot 0.00188 + 0.19^2 \cdot 0.0075 + 0.04^2 \cdot 0.041 = 0.00145,$$

and the standard deviation is $\hat{\sigma}_{V^\star} = 0.038$.

Combining the results with *equal* weights would produce a considerably less accurate estimate of variance equal to $\frac{1}{9}(0.00188 + 0.0075 + 0.041) = 0.005598$. The estimate of the standard deviation is $\overline{\sigma}_{V^\star} = 0.075$, twice as large as the previous estimate.

Let $Y$ be any random variable with mean $\mu$ and standard deviation $\sigma$. In most cases which will be considered in the context of measurements, we deal with *positive* random variables, such as concentration, weight, size and velocity. Assume, therefore, that $Y$ is a positive random variable.

An important characterization of $Y$ is its *coefficient of variation* defined as the ratio of $\sigma$ over $\mu$. We denote it as c.v.$(Y)$:

$$\text{c.v.}(Y) = \frac{\sigma}{\mu}. \tag{2.1.29}$$

Small c.v. means that the probability mass is closely concentrated around the mean value.

Suppose we have a random sample of size $n$ from population with mean $\mu$ and standard deviation $\sigma$. Consider now the sample mean $\overline{Y}$. As follows from (2.1.23), it has standard deviation equal to $\sigma/\sqrt{n}$ and mean $\mu$. Thus,

$$\text{c.v.}(\overline{Y}) = \frac{\text{c.v.}(Y)}{\sqrt{n}}. \tag{2.1.30}$$

The standard deviation of $\overline{Y}$ in a sample of size $n$ is $\sigma_{\hat{\mu}} = \sigma/\sqrt{n}$. It is called the *standard error of the mean*. The corresponding estimate is

$$s_{\hat{\mu}} = \frac{s}{\sqrt{n}}. \tag{2.1.31}$$

In words: the standard deviation of the sample mean equals the estimated population standard deviation divided by the square root of the sample size. For the apple data in Example 2.1.1, $s_{\hat{\mu}} = 2.8/\sqrt{20} = 0.63$.

## 2.2 How to Round off the Measurement Data

Nowadays, everybody uses calculators or computers for computations. They provide results with with at least 4 or 5 significant digits. Retaining all these digits in presenting the measurement results (e.g. estimates of means and standard deviations) might create a false impression of extremely accurate measurements. How many digits should be retained in presenting the results?

Let us adopt the following rules, as in Taylor (1997) and Rabinovich (2000):

1. Experimental uncertainties should always be rounded to two significant digits.

2. In writing the final result subject to uncertainty, the last significant figure of this result should coincide with the last significant digit of the uncertainty.

3. Any numbers to be used in subsequent calculations should normally retain at least one significant figure more than is finally justified.

For example, suppose that in calculating the specific weight, we have obtained the weight 54.452 g, the volume 5.453 cm$^3$, and the overall uncertainty of the specific weight (an estimate of standard deviation) was calculated as 0.002552.

First, following rule 1, we round off the uncertainty to 0.0026. After calculating the specific weight as $54.452/5.453 = 9.985696$, we keep only 5 significant digits (rule 2) and present the final result in the form: $9.9857 \pm 0.0026$.

Another important issue is the accuracy used in recording the original measurements. Suppose that a digital voltmeter has a discrete scale with step $h$. If it gives a reading of 12.36 V, $h = 0.01$.

When a measurement instrument is not of digital type, the readings are made from a scale with marks. Typically, we round off the measurement results to the nearest half of the interval between the adjacent scale marks. For example, a spring weight has a scale with marks at every 10 g, and we record the weight by rounding the reading to the nearest 5 g. In this case $h = 5$ g.

The value of $h$ depends on the measurement instrument's precision. The *desired precision* must depend on the spread of the measurement results in the population.

Suppose that the distribution of random weights is such that all weights lie between 998 and 1012 g. Weighing on a scale with step $h = 10$ g would result in reading the weight either as 1000 or 1010 g. This will introduce an error which is quite large relative to the standard deviation in the population.

In the extreme case when the instrument scale is too "rough", it might happen that all our readings are identical. Then our estimate of the population standard deviation will be zero, certainly an absurd result.

A practical rule for the desired relationship between the scale step $h$ and the population standard deviation $\sigma$ is that *the scale step should not exceed one half of the standard deviation in the population* (see Cochran and Cox 1957, Sect. 3.3).

Let us check our apple example. We have $h = 1$ gram. The estimated standard deviation is $s = 2.8$. Half of this is greater than $h$, and therefore the measurements device is accurate enough.

**Definition 2.2.1.** A measurement procedure with $\delta = s/h \geq 2$ is called *regular*.

A measurement scheme with $\delta < 0.5$ we call *special*. To put it simply, special measurements are made using a measuring instrument with scale step greater than two population standard deviations.

If a measurement is special, then the rules for computing the estimates of population mean and standard deviation must be modified. We will describe these rules in Chapter 8, for the case of sampling from a normal population. As a rule, we will deal only with regular measurements. It is characteristic of special measurements that all measurement results take on usually not more than two or three adjacent numerical values on the scale of the measurement instrument.

## 2.3 Normal Distribution in Measurements

Let us return to (1.1.4), which represents the measurement result as a random variable. We did not make any assumptions concerning the distribution of $Y$. In fact, to compute estimates of means and standard deviations we can manage without these assumptions. However, there are more delicate issues, such as identifying outliers, constructing confidence intervals and testing hypotheses which require assumptions regarding the distribution function of the random variables involved.

Most of our further analysis is based on the assumption that the random variable $Y$ follows the so-called normal distribution. Later we will give reasons for this assumption. The normal distribution is specified by two parameters, which we denote by $\mu$ and $\sigma$. If $Y$ has a normal distribution, we denote it by $Y \sim N(\mu, \sigma^2)$. Let us for convenience recall the basic facts regarding the normal distribution.

Let $Z \sim N(\mu, \sigma^2)$. Then the density function of $Z$ is

$$f_Z(v) = \frac{1}{\sqrt{2\pi}\sigma} e^{-(v-\mu)^2/2\sigma^2}, \quad -\infty < v < \infty; \tag{2.3.1}$$

$\mu$ may be any real number and $\sigma$ any positive number. The parameters $\mu$ and $\sigma$ have the following meaning for the normal distribution:

$$E[Z] = \mu, \ \mathrm{Var}[Z] = \sigma^2.$$

It is more convenient to work with the so-called *standard* normal distribution having $\mu = 0$ and $\sigma = 1$ (Fig. 2.3). So, if $Z_0 \sim N(0, 1)$, then

$$f_0(v) = \frac{1}{\sqrt{2\pi}} e^{-v^2/2}. \tag{2.3.2}$$

The cumulative probability function

$$P(Z_0 \leq t) = \int_{-\infty}^{t} f_0(v)dv = \Phi(t) \tag{2.3.3}$$

gives the area under the standard normal density from $-\infty$ to $t$. It is called the normalized Gauss function. (In European literature, the name Gauss—Laplace function is often used.) The table of this function is presented in Appendix A.

It is important to understand the relationship between $Z \sim N(\mu, \sigma^2)$ and $Z_0 \sim N(0, 1)$. The reader will recall that if $Y \sim N(\mu, \sigma^2)$, then the mean value of $Y$ equals $\mu$, and the variance of $Y$ equals $\sigma^2$. Now

$$Z = \mu + \sigma \cdot Z_0, \ \text{or} \ Z_0 = \frac{Z - \mu}{\sigma}. \tag{2.3.4}$$

In simple terms, the random variable with normal distribution $N(\mu, \sigma^2)$ is obtained from the "standard" random variable $Z_0 \sim N(0, 1)$ by a linear transformation: multiplication by standard deviation $\sigma$ and addition of $\mu$.

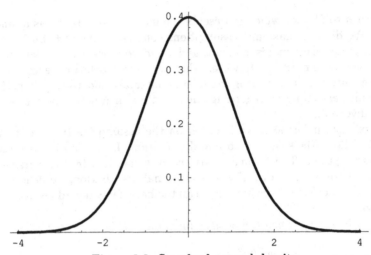

Figure 2.3. Standard normal density

For a better understanding of the normal density function, let us recall how the probability mass is distributed around the mean value. This is demonstrated by Fig. 2.4. In practical work with the normal distribution we often need to use so-called *quantiles*.

## Normal Quantiles

**Definition 2.2.1** Let $\alpha$ be any number between 0 and 1, and let $Y$ be a random variable with density $f_Y(t)$. Then the number $t_\alpha$ which satisfies the equality

$$\alpha = \int_{-\infty}^{t_\alpha} f_Y(t)dt \tag{2.3.5}$$

is called the $\alpha$-quantile of $Y$.

In simple terms, $t_\alpha$ is a number to the left of which lies the probability mass of size $\alpha$. Table 2.2 presents the most useful quantiles of the standard normal distribution. For this distribution, the $\alpha$-quantiles are usually denoted as $z_\alpha$. Note a useful property of these quantiles: for any $\alpha$,

$$z_{(1-\alpha)} = -z_\alpha. \tag{2.3.6}$$

How to calculate the quantile of an arbitrary normal distribution? Suppose $Y \sim N(\mu, \sigma^2)$. Then the $\alpha$-quantile of $Y$, denoted $t_\alpha^Y$ can be expressed via the standard normal quantile $z_\alpha$ as follows:

$$t_\alpha^Y = \mu + \sigma \cdot z_\alpha. \tag{2.3.7}$$

Table 2.2: Quantiles of the standard normal distribution

| $\alpha$ | $z_\alpha$ |
|-------|--------|
| 0.001 | −3.090 |
| 0.010 | −2.326 |
| 0.025 | −1.960 |
| 0.050 | −1.645 |
| 0.100 | −1.282 |
| 0.900 | 1.282 |
| 0.950 | 1.645 |
| 0.975 | 1.960 |
| 0.990 | 2.326 |
| 0.999 | 3.090 |

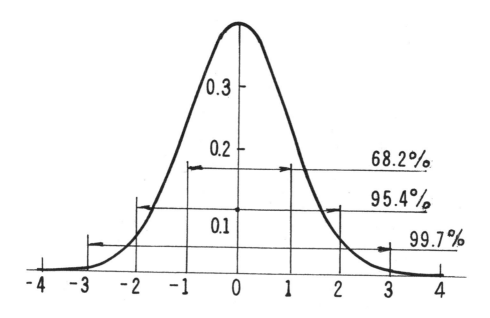

Figure 2.4. The probability mass under the normal density curve

## Normal Plot

A simple graphical tool can be used to check whether a random sample is taken from normal population. This is a so-called normal probability plot.

*Example 2.1.1 revisited* To construct the normal plot, we first have to order the weights, and assign to the $i$th ordered observation $y_{(i)}$ the value $p_{(i)} = (i-0.5)/n$; see Table 2.3. We now plot the points $(y_{(i)}, p_{(i)})$. Figure 2.5 shows the normal

Table 2.3: The ordered sample of 20 apples

| $i$ | Weight in grams | $p_{(i)}$ | $i$ | Weight in grams | $p_{(i)}$ |
|----|-----|-------|----|-----|-------|
| 1  | 158 | 0.025 | 11 | 166 | 0.525 |
| 2  | 162 | 0.075 | 12 | 166 | 0.575 |
| 3  | 163 | 0.125 | 13 | 167 | 0.625 |
| 4  | 163 | 0.175 | 14 | 167 | 0.675 |
| 5  | 164 | 0.225 | 15 | 167 | 0.725 |
| 6  | 164 | 0.275 | 16 | 168 | 0.775 |
| 7  | 165 | 0.325 | 17 | 168 | 0.825 |
| 8  | 165 | 0.375 | 18 | 169 | 0.875 |
| 9  | 165 | 0.425 | 19 | 169 | 0.925 |
| 10 | 166 | 0.475 | 20 | 170 | 0.975 |

plot for the apple weights. If the points on the plot form a reasonable straight line, then this is considered as a confirmation of the normality for the population from which the sample has been taken.

### The Origin of the Normal Distribution

To explain the use of the normal distribution in measurements, let us quote Box and Luceno (1997, p. 51).

"Why should random errors tend to have a distribution that is approximated by the normal rather than by some other distribution? There are many different explanations – one of which relies on the central limit effect. This central limit effect does not apply only to averages. If the overall error $e$ is an aggregate of a number of component errors $e = \epsilon_1 + \epsilon_2 + \ldots + \epsilon_n$, such as sampling errors, measurement errors or manufacturing variations, in which no one component dominates, then almost irrespective of the distribution of the individual components, it can be shown that the distribution of the aggregated error $e$ will tend to the normal as the number of components gets larger".

A different and very remarkable argument is due to James Clark Maxwell. We describe it following Krylov (1950, Sect. 121) and Box and Luceno (1997). Suppose that we perform a large number $N$ of measurements, each of which is carried out with a random error $\epsilon$. Let us consider the number of times (out of $N$) that the random error falls into a small interval $[x, x + dx]$. It is natural to assume that this number is proportional to $dx$; proportional to $N$; and dependent on $x$. Denoting this number by $ds$, we have

$$ds = Nf(x)dx. \qquad (2.3.8)$$

Thus the probability that the random error is between $x$ and $x + dx$ is

$$dp = f(x)dx. \qquad (2.3.9)$$

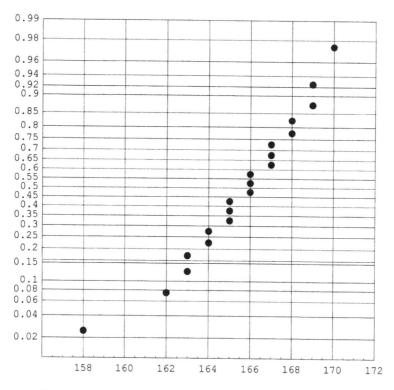

Figure 2.5. The normal plot for the data of Example 2.1.1

The function $f(x)$ is what we used to call the density function. Now it is natural to postulate that this function is even, i.e. $f(x) = f(-x)$, which means that the distribution of errors is symmetric around the origin $x = 0$. Without loss of generality, we may assume that $f(x) = \phi(x^2)$.

The second postulate is that $\phi(\cdot)$ is a decreasing function of $x$, i.e. larger errors are less probable than smaller ones. Suppose now that we carry out a measurement whose result is a pair of numbers $(x, y)$. In other words, let us assume that we have a two-dimensional measurement error $(\epsilon_1, \epsilon_2)$. A good way to imagine this is to consider shooting at a target, and interpret $\epsilon_1$ and $\epsilon_2$ as the horizontal and vertical deviations from the center of the target, respectively. Put the center of the target at the point $(x = 0, y = 0)$.

The third postulate is the following: the random errors $\epsilon_1$ and $\epsilon_2$ are independent. This means that the probability $dp$ of hitting a target at a small square $[x, x + dx] \times [y, y + dy]$ can be represented in the following product form:

$$dp = \phi(x^2)\phi(y^2)dxdy. \tag{2.3.10}$$

Now rotate the target around the origin in such a way that the $x$-axis goes through the point $(x, y)$. Then it can be shown, based on the third postulate, that the probability of hitting the target at the distance $r = \sqrt{x^2 + y^2}$ from the origin depends on $x$ and $y$ through $r$ only. In other words, the joint density of $\epsilon_1, \epsilon_2$ must depend only on $r = \sqrt{x^2 + y^2}$. Thus we arrive at the following principal relationship:

$$\phi(x^2)\phi(y^2) = C \cdot \phi(x^2 + y^2). \tag{2.3.11}$$

The solution of (2.3.11) is beyond the standard calculus course. Let us accept without proof the fact that (2.3.11) determines a unique function which has the following general form: $\phi(x^2) = C \cdot \exp^{Ax^2}$. Since $\phi(\cdot)$ must be decreasing (the second postulate), $A$ is negative. Denote it by $-1/2\sigma^2$. The constant $C$ must be chosen such that the integral of $\phi$ from minus to plus infinity is 1. It turns out that $C = 1/\sqrt{2\pi}\sigma$. Thus we arrive at the expression

$$f(x) = \phi(x^2) = \frac{1}{\sqrt{2\pi}\sigma} e^{-x^2/2\sigma^2}, \tag{2.3.12}$$

which is the normal density centered at zero.

## 2.4   The Uniform Distribution

The second distribution widely used in measurements is called *the rectangular* or *uniform* distribution.

The density of this distribution has a rectangular form:

$$f_X(x) = \frac{1}{b-a}, \; x \in [a, b], \tag{2.4.1}$$

with $f(x) = 0$ outside $[a, b]$. The mean value of $X$ is $E[X] = (b + a)/2$ and the variance is $\mathrm{Var}[X] = (b - a)^2/12$.

Often the interval $[a, b]$ is symmetric with respect to zero: $[a, b] = [-\Delta, \Delta]$, i.e. the density is constant and differs from zero in a symmetric interval around zero. The length of this interval is $2\Delta$. We use the shorthand notation $X \sim U(a, b)$ to say that X has density (2.4.1).

Let us recall that the first two moments of $X \sim U(-\Delta, \Delta)$ are

$$E[X] = 0, \tag{2.4.2}$$

$$\mathrm{Var}[X] = \frac{\Delta^2}{3}. \tag{2.4.3}$$

The standard deviation is

$$\sigma_X = \frac{\Delta}{\sqrt{3}}. \tag{2.4.4}$$

Now suppose that the random variable $X$ is a sum of several independent random variables $X_1, X_2, \ldots, X_k$, each of which has a uniform distribution:

Table 2.4: Atomic weights

| Element | Atomic weight | Quoted Uncertainty | Standard Uncertainty |
|---------|---------------|--------------------|----------------------|
| C | 12.011 | ±0.001 | 0.00058 |
| H | 1.00794 | ±0.00007 | 0.000040 |
| O | 15.9994 | ±0.0003 | 0.0017 |
| K | 39.0983 | ±0.001 | 0.00058 |

$X_i \sim U(-\Delta_i, \Delta_i)$. The variance of sum of independent random variables is the sum of component variances. Thus,

$$\text{Var}[X] = \sum_{i=1}^{k} \frac{4\Delta_i^2}{12} = \sum_{i=1}^{k} \frac{\Delta_i^2}{3}. \tag{2.4.5}$$

The standard deviation of $X$ is

$$\sigma_X = \sqrt{(\Delta_1^2 + \ldots + \Delta_k^2)/3}. \tag{2.4.6}$$

An important and useful fact is that if the number of summands $k$ is 6 or more and the $\Delta_i$ are of the same magnitude, then the distribution of $X$ may be quite well approximated by a normal distribution. So, for $k \geq 6$,

$$X \sim N(0, \sigma_X^2), \tag{2.4.7}$$

where $\sigma_X$ is given by (2.4.6).

Let us consider several typical cases of the use of the uniform distribution in measurement practice.

*Example 2.4.1: Round-off errors*
Suppose that the scale of a manometer is graduated in units of 0.1 atmosphere. The pressure reading is rounded therefore to the closest multiple of 0.1. This introduces a round-off error which may be assumed to have a uniform distribution in the interval $[-0.05, 0.05]$.

*Example 2.4.2: Atomic weights*
Table 2.4 presents the atomic weights of four elements, together with the uncertainties in their measurements. Eurachem (2000) explains for the data of Table 2.4 that "for each element, the standard uncertainty is found by treating the quoted uncertainty as forming the bounds of a rectangular distribution. The corresponding standard uncertainty is therefore obtained by dividing these values by $\sqrt{3}$" (see (2.4.4)).

*Example 2.4.3: Specification limits* Eurachem (2000) describes how to handle errors arising due to the geometric imperfection of chemical glassware instruments. For example, the manufacturer of a 10 ml pipette guarantees that the volume of the liquid in the pipette filled up to the 10 ml mark with "absolute" accuracy may deviate from this value by not more than ±0.012 ml. In simple terms, the error in the liquid volume due to the geometric imperfection of the glassware is $X \sim U(-0.012, 0.012)$.

## 2.5   Dixon's Test for a Single Outlier

Looking at the box and whisker plot in Fig. 2.2, we see that the smallest observation $y_{(1)} = 158$ may possibly be an outlier.   Below we describe a simple test due to Dixon; see Bolshev and Smirnov (1965, p. 89). The test is used to identify a single outlier in a sample from a normal population.

Suppose we observe a random sample $\{y_1, \ldots, y_n\}$. Denote the $k$th ordered observation as $y_{(k)}$. Then the smallest and the largest observations are denoted as $y_{(1)}$ and $y_{(n)}$, respectively.   Our null hypothesis $\mathcal{H}_0$ is the following:  all observations belong to a normal population with unknown parameters $(\mu, \sigma^2)$. The alternative hypothesis $\mathcal{H}_1^-$ is that the smallest observation comes from a population with distribution $N(\mu - d, \sigma^2)$, where $d$ is some unknown positive constant. The test statistic $Q$ is the following:

$$Q = \frac{y_{(2)} - y_{(1)}}{y_{(n)} - y_{(1)}}. \tag{2.5.1}$$

The difference $y_{(n)} - y_{(1)}$ is called the observed or sample range.

If $Q$ exceeds the critical value $C_n$ given in Table 2.5, $\mathcal{H}_0$ is rejected (at significance $\alpha = 0.05$) in favor of the alternative $\mathcal{H}_1^-$. Let us apply Dixon's test to the apple data. The value of $Q$ is $Q = (162 - 158)/(170 - 158) = 0.333$. It is greater than the critical value $C_n = 0.300$ for $n = 20$. Thus, we assume that the value $y_{(1)} = 158$ is an outlier.

For the alternative $\mathcal{H}_1^+$ that the largest observation comes from a population $N(\mu + d, \sigma^2)$, where $d > 0$, we use a similar statistic

$$Q^\star = \frac{y_{(n)} - y_{(n-1)}}{y_{(n)} - y_{(1)}}. \tag{2.5.2}$$

We reject the null hypothesis in favor of $\mathcal{H}_1^+$, at significance level $\alpha = 0.05$, if $Q^\star$ exceeds the critical value $C_n$ in Table 2.5.

*Remark 1: Interpretation of outliers*
The easiest decision is just to delete the outlying observation from the sample. The right way to deal with an outlier is to try to find the reason for its appearance. It might be an error in data recording, some fault of the measuring device or some reason related to the production process itself. The latter is most important for the process study and should deserve the greatest attention of the statistician.

Table 2.5: Dixon's statistic for significance level $\alpha = 0.05$ (Bolshev and Smirnov, p. 328)

| $n$ | $C_n$ | $n$ | $C_n$ |
|---|---|---|---|
| 3 | 0.941 | 13 | 0.361 |
| 4 | 0.765 | 14 | 0.349 |
| 5 | 0.642 | 15 | 0.338 |
| 6 | 0.560 | 16 | 0.329 |
| 7 | 0.507 | 17 | 0.320 |
| 8 | 0.468 | 18 | 0.313 |
| 9 | 0.437 | 19 | 0.306 |
| 10 | 0.412 | 20 | 0.300 |
| 11 | 0.392 | 25 | 0.277 |
| 12 | 0.376 | 30 | 0.260 |

It is interesting to check the normal plot without the outlier $y = 158$. The normal plot without the "outlier" in Fig. 2.6 looks "more normal".

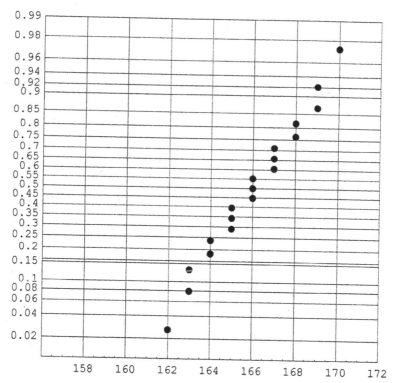

Figure 2.6. The normal plot for 19 observations without the outlier

Table 2.6: Coefficients for computing Estimates of $\sigma$ based on sample range

| $n$ | $A_n$ | $u_n$ |
|---|---|---|
| 2 | 1.128 | 1.155 |
| 3 | 1.693 | 1.732 |
| 4 | 2.059 | 2.078 |
| 5 | 2.326 | 2.309 |
| 6 | 2.534 | 2.474 |
| 7 | 2.704 | 2.598 |
| 8 | 2.847 | 2.694 |
| 9 | 2.970 | 2.771 |
| 10 | 3.078 | 2.834 |
| 12 | 3.258 | 2.931 |
| 14 | 3.407 | 3.002 |
| 16 | 3.532 | 3.057 |
| 18 | 3.640 | 3.099 |
| 19 | 3.689 | 3.118 |
| 20 | 3.735 | 3.134 |

## 2.6  Using Range to Estimate $\sigma$

If a sample of size $n \leq 10$ is drawn from a normal population, then its *range*, i.e. the largest observation minus the smallest one, can serve as a quick and quite accurate estimate of $\sigma$. The *modus operandi* is very simple: to obtain the estimate of $\sigma$, sample range must be divided by the coefficient $A_n$ taken from Table 2.6:

$$\hat{\sigma} = \frac{\text{sample range}}{A_n}. \tag{2.6.1}$$

Let us apply (2.6.1) to the Yarok example, after deleting the outlier $y_{(1)} = 158$. Using Table 2.6, we have $\hat{\sigma} = 8/A_{19} = 2.17$. Note that the estimate of $\sigma$ based on 19 observations computed by (2.1.7) is $s = 2.24$, close enough to the previous result.

### Properties of $\sigma$-estimators

Suppose that our observations $\{y_1, \ldots, y_n\}$ are drawn from a uniform population $Y \sim U(a, b)$. It turns out also for this case that there is an easy way to estimate the standard deviation using sample range.

It is easily proved that the mean value of the range for a sample drawn from a uniform population equals

$$E[Y_{(n)} - Y_{(1)}] = (b - a)(n - 1)/(n + 1). \tag{2.6.2}$$

Compare this formula with the formula for the standard deviation:

$$\sigma_Y = (b - a)/\sqrt{12} \tag{2.6.3}$$

and derive that

$$\sigma_Y = E[Y_{(n)} - Y_{(1)}](n + 1)/(\sqrt{12}(n - 1)). \tag{2.6.4}$$

It follows therefore that an estimate of $\sigma_Y$ can be obtained by the formula

$$\hat{\sigma}_Y = \frac{y_{(n)} - y_{(1)}}{u_n}, \tag{2.6.5}$$

where $u_n = \sqrt{12}(n - 1)/(n + 1)$.

The coefficients $u_n$ are given in Table 2.6. It is surprising that for small $n$ they are quite close to the values $A_n$. This means that for estimating $\sigma$ in small samples ($n \leq 10$), assuming a normal distribution or uniform distribution would produce quite similar results.

We see that there are two different estimators of standard deviation: the $S$-estimator,

$$S = \sqrt{\frac{\sum_{i=1}^{n}(Y_i - \overline{Y})^2}{n - 1}}, \tag{2.6.6}$$

see (2.1.7), and the estimator (2.6.1) based on range,

$$\hat{\sigma} = \frac{Y_{(n)} - Y_{(1)}}{A_n}.$$

The latter is recommended for a normal sample, when the coefficients $A_n$ are taken from Table 2.6, while the former one is universal and does not need any distributional assumptions regarding the random variables involved. Suppose that we may assume normality. Which of the above two estimators is preferable?

Let us note first that the estimator $S$ is *biased*, i.e. its mean value is not equal to $\sigma$. To make this estimator unbiased it must be multiplied by a coefficient

$$b_n = \frac{\Gamma((n - 1)/2)\sqrt{n - 1}}{\Gamma(n/2)\sqrt{2}}, \tag{2.6.7}$$

where $\Gamma(\cdot)$ is the gamma function. More details can be found in Bolshev and Smirnov (1965, p. 60). The values of $b_n$ are given in Table 2.7.

Suppose that the $S$-estimator is multiplied by $b_n$. Let us now compare $S^{\star} = b_n S$ with $\hat{\sigma}$ (2.6.1) Both formulas define unbiased estimators. They can be compared in terms of their variance. The ratio $e_n = \text{Var}[S^{\star}]/\text{Var}[\hat{\sigma}]$ is called *efficiency*. Bolvev and Smirnov (1965, p. 60), present the following values of $e_n$: $e_2 = 1, e_5 = 0.96, e_{10} = 0.86, e_{15} = 0.77$. We see that for small $n \leq 10$, the use of $\hat{\sigma}$ leads to a rather small loss of efficiency, i.e. $\text{Var}[\hat{\sigma}]$ is only slightly larger than $\text{Var}[S^{\star}]$. This fact justifies the use of the estimator based on range in the case of small samples.

In addition, let us mention the theoretical fact that for the normal case, the $S^{\star}$-estimator is the *minimal-variance* unbiased estimator, i.e. it has the smallest possible variance among all possible unbiased estimators of $\sigma$.

Table 2.7: The correction factor $b_n$

| $n$ | $b_n$ |
|----|-------|
| 2  | 1.253 |
| 3  | 1.128 |
| 4  | 1.085 |
| 5  | 1.064 |
| 10 | 1.028 |
| 15 | 1.018 |
| 20 | 1.013 |
| 30 | 1.009 |
| 50 | 1.005 |

Table 2.8: $\gamma_n(\alpha) = t_\alpha(n-1)/\sqrt{n}$ for $\alpha = 0.05$ and $\alpha = 0.025$

| Sample size $n$ | $\gamma_n(0.05)$ | $\gamma_n(0.025)$ | Sample size $n$ | $\gamma_n(0.05)$ | $\gamma_n(0.025)$ |
|------|-------|-------|------|-------|-------|
| 3  | 1.686 | 2.484 | 13 | 0.494 | 0.604 |
| 4  | 1.176 | 1.591 | 14 | 0.473 | 0.577 |
| 5  | 0.953 | 1.241 | 15 | 0.455 | 0.554 |
| 6  | 0.823 | 1.050 | 16 | 0.438 | 0.533 |
| 7  | 0.734 | 0.925 | 17 | 0.423 | 0.514 |
| 8  | 0.700 | 0.836 | 18 | 0.410 | 0.497 |
| 9  | 0.620 | 0.769 | 19 | 0.398 | 0.482 |
| 10 | 0.580 | 0.715 | 20 | 0.387 | 0.468 |
| 11 | 0.546 | 0.672 | 25 | 0.342 | 0.413 |
| 12 | 0.518 | 0.635 | 30 | 0.310 | 0.373 |

## 2.7   Confidence Interval for the Population Mean

Suppose we draw a sample of size $n$ from a normal population. It is well known that the $1 - 2\alpha$ confidence interval for the population mean $\mu$ is the following:

$$\left(\bar{y} - t_\alpha(n-1)\frac{s}{\sqrt{n}}, \bar{y} + t_\alpha(n-1)\frac{s}{\sqrt{n}}\right), \tag{2.7.1}$$

where $t_\alpha(n-1)$ is the $\alpha$-critical value of the $t$-statistic with $n-1$ degrees of freedom. To simplify the computation procedure, we present in Table 2.8 the values of $\gamma_n(\alpha) = t_\alpha(n-1)/\sqrt{n}$ for various values of $n$ and for $\alpha$ equal to 0.05 and 0.025.

For the Yarok example, after deleting the outlier, the sample mean is 166.00, $n = 19$, $s = 2.24$. Let us keep two significant digits for $s$ (put $s = 2.2$) and

Table 2.9: Measurement results for ten specimens

| $i$ | $D_i$ | $Y_i$ | $\Delta_i = Y_i - D_i$ |
|-----|-------|-------|------------------------|
| 1 | 10.533 | 10.545 | 0.012 |
| 2 | 9.472 | 9.476 | 0.004 |
| 3 | 9.953 | 9.960 | 0.007 |
| 4 | 10.823 | 10.830 | 0.007 |
| 5 | 8.734 | 8.736 | 0.002 |
| 6 | 10.700 | 10.706 | 0.006 |
| 7 | 9.620 | 9.630 | 0.010 |
| 8 | 10.580 | 10.582 | 0.002 |
| 9 | 10.546 | 10.555 | 0.011 |
| 10 | 9.518 | 9.520 | 0.002 |

round off the final computation result to 0.1. Thus, using Table 2.8, for $n = 19$, and $1 - 2\alpha = 0.9$, we have $\gamma_{19}(0.05) = 0.398$. Thus the 0.90 confidence interval on the sample mean is

$$[166.00 - 0.398 \cdot 2.2, 166.00 + 0.398 \cdot 2.2] = [165.1, 166.9]$$

Let us recall that confidence intervals can be used to test a hypothesis regarding the mean value in the following way. Suppose that we want to test the null hypothesis $\mathcal{H}_0$ that the population mean equals $\mu = \mu_0$ against a two-sided alternative $\mu \neq \mu_0$, at the significance level 0.05. We proceed as follows. Construct, on the basis of the data, a 0.95 confidence interval on $\mu$. If this interval covers the value $\mu_0$, the null hypothesis is *not* rejected. Otherwise, it is rejected at the significance level 0.05.

*Remark 1*
We draw the reader's attention to an important issue which is typically not very well stressed in courses on statistics. If the sample $\{y_1, \ldots, y_n\}$ is obtained as a result of measurements with a *constant* but unknown bias $\delta$, then the confidence interval (2.7.1) will be in fact a confidence interval for the sum $\mu + \delta$.

*Example 2.7.1: Testing a measurement instrument for bias*
In order to test whether there is a systematic error in measuring the diameter of a cylindric shaft by a workshop micrometer, ten specimens were manufactured. Each was measured twice. The first measurement was made by a highly accurate "master" instrument in the laboratory. This instrument is assumed to have a zero bias. The second measurement was made by the workshop micrometer.

Denote by $D_i$ and $Y_i$ the measurement results for the $i$th specimen made by the master and the workshop micrometer, respectively. The measurement results are presented in Table 2.9. The average value of the difference $\Delta_i =$

$Y_i - D_i$ is 0.0063, and the sample standard deviation for $\Delta_i$ is $s = 0.0038$. If the workshop micrometer has zero bias, the mean of $\Delta_i$ must be zero. So, our null hypothesis $\mathcal{H}_0$ is $E[\Delta_i] = \mu_\Delta = 0$.

Let us construct a 0.9 confidence interval on the mean value $\mu$ of the differences $\Delta_i$. From Table 2.8, $\gamma_{10} = 0.580$. Thus the confidence interval is

$$[0.0063 - 0.580 \times 0.0038, 0.0063 + 0.580 \times 0.0038] = [0.0041, 0.0085].$$

This interval does not contain zero, and thus the null hypothesis is rejected. The estimate of the bias of the workshop micrometer is $\overline{\mu}_\Delta = 0.0063$. In words: on average, the workshop micrometer measurements are larger than the true value by approximately 0.0063 mm.

The probability model for the above measurement scheme is as follows. $D_i$ is considered as the true, "absolutely accurate" value of the diameter of the $i$th specimen. The measurement result is given by

$$Y_i = D_i + \mu_\Delta + \epsilon_i, \tag{2.7.2}$$

where $\mu_\Delta$ is the bias of the workshop micrometer, and $\epsilon_i$ is the random zero-mean measurement error. By our assumption, therefore, the workshop micrometer has a constant bias during the measurement experiment.

## 2.8   Control Charts for Measurement Data

Control charts were first invented by W.A. Shewhart in the 1930s for monitoring production processes and for discovering deviations from their normal course. These deviations in their most basic form are of two types – changes in the process mean value and changes in the standard deviation. By a "production process" we mean here an one-dimensional parameter, say the diameter of a rod produced on an automatic machine.

In a normal, stable situation the rod diameter variations are described by a random variable $Y$ with mean $\mu$ and standard deviation $\sigma$. These variations are caused by so-called common causes. Beauregard et al. (1992, p. 15) say that the "common cause of variation refers to chronic variation that seems to be 'built-in' to the process. It's always been there and it will always be there unless we change the process".

Shifts in the mean value and changes in the standard deviation result from the action of so-called special causes, such as change of adjustment, deviation in the technological process, sudden changes in the measurement process etc. According to Beauregard et al., "special causes of variation are due to acute or short-term influences that are not normally a part of the process as it was intended to operate".

We explain the Shewhart chart for measurement data using an example borrowed from Box and Luceno (1997, p. 57).

Table 2.10 shows measurements of the diameter of rods taken from production line. The target value of the diameter is $T = 0.876$ inch. A sample of four

Table 2.10: Rod diameter measurements over 20 hours

| Hour $i$ | $D_1$ | $D_2$ | $D_3$ | $D_4$ | Average | Range |
|---|---|---|---|---|---|---|
| 1 | −8 | 3 | 12 | −9 | −0.50 | 21 |
| 2 | −3 | 7 | −9 | 20 | 3.75 | 29 |
| 3 | 7 | −5 | 7 | −8 | 0.25 | 15 |
| 4 | 18 | −23 | −1 | −18 | −6.00 | 41 |
| 5 | 0 | 5 | 10 | −11 | 1.00 | 21 |
| 6 | 1 | −9 | 5 | 9 | 1.50 | 18 |
| 7 | 5 | −2 | 25 | −3 | 6.25 | 28 |
| 8 | 7 | 4 | −11 | −14 | −3.50 | 21 |
| 9 | 9 | −19 | 7 | 19 | 4.00 | 38 |
| 10 | 11 | 28 | 11 | 7 | 14.25 | 21 |
| 11 | 0 | 0 | 5 | 12 | 4.25 | 12 |
| 12 | 4 | −6 | 14 | 5 | 4.25 | 20 |
| 13 | 2 | 1 | 21 | 0 | 6.00 | 21 |
| 14 | 28 | 13 | −9 | −4 | 7.00 | 37 |
| 15 | 26 | 2 | 7 | −10 | 6.25 | 36 |
| 16 | −15 | 11 | 0 | 7 | 0.75 | 26 |
| 17 | 14 | 16 | 7 | 23 | 15.00 | 16 |
| 18 | 19 | 11 | 29 | 18 | 19.25 | 18 |
| 19 | 16 | 20 | 30 | 17 | 20.75 | 14 |
| 20 | −4 | 24 | 9 | 26 | 13.75 | 30 |

rods was randomly chosen and measured every hour and the table shows data over a particular period of 20 hours of production. The figures in this table are the deviations of the rod diameters from value $T_1 = 0.870$ in thousands of an inch. Each row in the table corresponds to one hour. The last two columns give the hour average and hour range for the four observations.[1]

Two graphs are of the utmost importance in Shewhart's methodology. The first, the "$X$ bar chart" (see Fig. 2.7), is a representation of *sample means* for each hour.

---

[1] This material is borrowed from George Box and Alberto Luceno *Statistical Control by Monitoring and Feedback Adjustment* (1997) and is used by permission of John Wiley & Sons, Inc., Copyright ©1997.

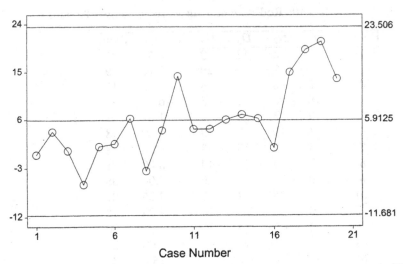

Figure 2.7. The $X$ bar chart for data in Table 2.10 produced by Statistix. The zero line corresponds to $T_1 = 0.870$.

### $X$ Bar Chart

The line 5.9125 (near the mark "6") is the estimated mean value of the process. This value is the average $\hat{\mu}$ of hour averages in Table 2.10. The lines 23.506 and $-11.681$ (near the marks "24" and "12", respectively) are called control limits and correspond to $\pm 3\hat{\sigma}_D/\sqrt{4}$ deviations from the estimated mean value. $\hat{\sigma}_D$ is the estimate of the standard deviation of the rod diameter. We divide it by $\sqrt{4}$ because we are plotting $\pm 3$ standard deviations for the average of 4 observations.

An important statistical fact is that the sample means, due to the central limit effect, closely follow the normal distribution, even if the rod diameters themselves may not be normally distributed.

How do we estimate the standard deviation $\hat{\sigma}_D$? The practice is to use the *average range*. From the *Range* column of Table 2.10 one can calculate that the average range is $\overline{R} = 24.15$. Now from Table 2.6 obtain the coefficient $A_n$ for the sample size $n = 4$: $A_4 = 2.059$. Then, as already explained in Sect. 2.6, the estimate of $\sigma_D$ will be

$$\hat{\sigma}_D = \overline{R}/A_4 = 24.15/2.059 = 11.73. \qquad (2.8.1)$$

The estimate of standard deviation of sample averages is $\sqrt{4} = 2$ times smaller and equals 5.86. We can round this result to 5.9.

When the averages go outside of the control limits, actions must be taken to establish the special cause for that.

The principal assumption of Shewhart's charts is that the observations taken each hour are, for a state of control, sample values of identically distributed independent random variables. The deviations from these assumptions manifest themselves in the appearance of abnormalities in the behaviour of the $X$ bar plot.

For example, Western Electric Company adopted the following rules for an action signal (see Box and Luceno (1997)):

Rule 1: A single point lies beyond the three-sigma limits.

Rule 2: Two out of three consecutive points lie beyond the two-sigma limits.

Rule 3: Four out of five consecutive points lie beyond one-sigma limits.

Rule 4: Eight consecutive points lie on one side of the target value.

Nelson (1984) contains a collection of stopping rules based on various abnormal patterns in the behavior of $X$ bar charts.

## Range charts

The other graph which is important in Shewhart's methodology is the 'range chart' or '$R$ chart'. Range charts are meant for detecting changes in the process variability. They are constructed similarly to the $X$ bar charts; see Fig. 2.8. Each hour, the actual observed range of the sample of $n = 4$ rod diameters is plotted on the chart.

Denote by $\sigma_R$ the estimate of the standard deviation of the range (for $n = 4$). The distribution of the range is only approximately normal, but nevertheless $\pm 2\sigma_R$ or $\pm 3\sigma_R$ limits are used to supply action limits and warning limits.

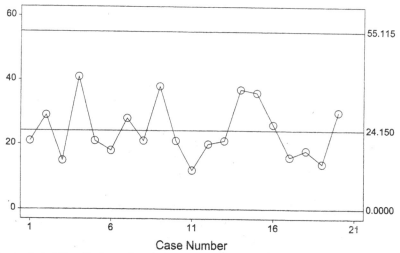

Figure 2.8. The $R$ chart for the data in Table 2.10. The average range is $\bar{r} = 24.15$. The upper control limit equals $55.115 = 24.150 + 3 \cdot 10.32$.

For the normal distribution, the mean range and the standard deviation of the range (for fixed sample size $n$) depend only on $\sigma$ and are proportional to $\sigma$,

Table 2.11: Probabilistic characteristic of normal ranges, $\sigma = 1$; Bolshev and Smirnov (1965), p. 58

| $n$ | $E[R_n]$ | $\sigma_{R_n}$ | $k_R(n)$ | $UCL = E[R_n] + 3\sigma_{R_n}$ | $P_0 = P(R_n > UCL)$ |
|-----|----------|----------------|----------|-------------------------------|----------------------|
| 4   | 2.059    | 0.880          | 2.34     | 4.699                         | 0.0049               |
| 5   | 2.326    | 0.864          | 2.69     | 4.918                         | 0.0046               |
| 6   | 2.534    | 0.848          | 2.99     | 5.078                         | 0.0045               |
| 7   | 2.704    | 0.833          | 3.25     | 5.203                         | 0.0044               |
| 8   | 2.847    | 0.820          | 3.47     | 5.307                         | 0.0043               |
| 9   | 2.970    | 0.808          | 3.68     | 5.394                         | 0.0044               |
| 10  | 3.078    | 0.797          | 3.86     | 5.469                         | 0.0044               |

the standard deviation of the population. $\sigma_R$ can be obtained from the average range $\bar{r}$ by the following formula:

$$\hat{\sigma}_R = \frac{\bar{r}}{k_R(n)}, \tag{2.8.2}$$

where $k_R(n)$ is the coefficient in the fourth column of Table 2.11.

In our example, the average of the last column of Table 2.10 is 24.15. $k_R(4) = 2.34$ and

$$\hat{\sigma}_R = \frac{24.15}{2.34} = 10.32. \tag{2.8.3}$$

If the average range minus $3\hat{\sigma}_R$ goes below zero, it is replaced by zero.

### Influence of Measurement Errors on the Performance of $R$ Charts

The above description of Shewhart's charts is quite basic. There is a huge literature devoted to the control and monitoring of changes in the process which might be revealed by these charts; see Box and Luceno (1997) and Beauregard et al. (1992). In practice, it is most important to find the so-called assignable causes responsible for the particular deviation beyond the control limits on the $X$ bar and/or the $R$ chart.

The literature on statistical process control typically devotes little attention to the fact that the measurement results reflect both the drift and the variability of the controlled process *and* the systematic and random measurement errors. Formally, the measurement result $Y$ might be represented in the following form:

$$Y = \mu + \delta_{pr} + \epsilon_{pr} + \delta_{instr} + \epsilon_{instr}, \tag{2.8.4}$$

where $\mu$ is the process target (mean) value, $\delta_{pr}$ is the shift of the process from $\mu$, $\epsilon_{pr}$ is the random deviation of the process around $\mu + \delta_{pr}$ due to common causes and $\delta_{instr}$ and $\epsilon_{instr}$ are the systematic bias and random error of the measurement instrument, respectively.

Suppose that statistical process control is applied to a process with $\delta_{pr} = 0$ and $\delta_{instr} = 0$, i.e. the process has no shift and the measurement instrument has no systematic bias. Then

$$Y = \mu + \epsilon_{pr} + \epsilon_{instr}. \tag{2.8.5}$$

Assuming that $\epsilon_{pr}$ and $\epsilon_{instr}$ are independent, we have

$$\sigma_Y = \sqrt{\sigma_{pr}^2 + \sigma_m^2}, \tag{2.8.6}$$

where $\sigma_{pr}$ and $\sigma_m$ are the standard deviations of $\epsilon_{pr}$ and $\epsilon_{instr}$, respectively.

If for example $\gamma = \sigma_m/\sigma_{pr} = 1/7$, then $\sigma_Y = \sqrt{\sigma_{pr}^2 + \sigma_{pr}^2/49} \approx 1.01\sigma_{pr}$. Thus the overall standard deviation will increase by 1 % only. Assuming that 1 % increase is admissible, we can formulate the following rule of thumb: The standard deviation of the random measurement error must not exceed one seventh of the standard deviation caused by the common causes in the controlled process.

There are situations, especially in chemical measurements, for which typically the measurement error has the same magnitude as the process variation; see Dechert et al. (2000). These situations demand special attention since the presence of large measurement errors may change the performance characteristics of the control charts.

Suppose that the $X$ bar and the $R$ charts were designed using properly adjusted and highly accurate measurement instruments and/or measurement procedures. In other words, let us assume that the control limits were established for a process in control, where the measurement instruments had a negligible $\delta_{instr}$ and $\sigma_m$. Afterwards, in the course of work, the measurement instrument gradually loses its proper adjustment and develops a systematic error, say a positive bias $\delta > 0$. Then, even in the absence of any special cause, the $X$ bar chart will reveal a high number of upper control limit crossings. Similarly, an increase in $\sigma_m$, e.g. due to measurement instrument mechanical wearout, will lead to increase in the probability of crossing the control limits, both for $X$ bar and $R$ charts. It should be kept in mind, therefore, that the loss of accuracy of a measurement instrument may be the assignable cause for the crossing of control limits in Shewhart's charts.

Note also that the appearance of instrument bias in measurements will not affect the range of the sample, and thus the $R$ chart will be insensitive ("robust") to the systematic error in the measurement process. If the $R$ chart looks "normal", but the $X$ bar chart signals trouble, a reasonable explanation might be that this trouble is caused by a process drift or by measurement instrument bias.

Any measurement instrument and/or measurement procedure based on using measurement instruments must be subject to periodic calibration, i.e. inspections and check-ups using more accurate and precise instruments. This calibration reminds in principle the statistical process control. In particular, the measurement instruments must be checked periodically to discover changes in the adjustment (e.g. the presence of a systematic bias) and to discover a loss of precision, i.e. an increase in the standard deviation of random measurement errors. The particular details how to carry out these inspections depend on the type of the instrument, its precision standards, etc.; see Morris (1997).

Let us investigate the influence of measurement errors on the performance of the $R$ chart. This material is of somewhat theoretical in nature and the formal reasoning could be omitted at the first reading. We assume that the process measurement model is the following:

$$Y = \mu + \epsilon_{pr} + \epsilon_{instr}, \tag{2.8.7}$$

where $\epsilon_{pr} \sim N(0, \sigma_{pr}^2)$ and $\epsilon_{instr} \sim N(0, \sigma_m^2)$ are independent. Expression (2.8.7) means that we assume no systematic process shift and no measurement bias: $\delta_{pr} = \delta_{instr} = 0$. In fact, this assumption is not restrictive since $R$ bar charts are insensitive to process and measurement bias.

Let us consider the probability of crossing the upper control limit in the $R$ chart:

$$P_0 = P(R_n > E[R_n] + 3\sigma_{R_n}),$$

where $R_n$ is the random range in the sample of $n$ normally distributed measurement results, $E[R_n]$ is the corresponding mean range and $\sigma_{R_n}$ is the range standard deviation.

The most important characteristic of the performance of a control chart is the average run length (ARL), defined as the mean number of samples until the first crossing of the three-sigma control limit. ARL is expressed in terms of $P_0$ as follows:

$$\text{ARL} = \frac{1}{P_0}. \tag{2.8.8}$$

To compute $P_0$ note that $R_n/\sigma$ is distributed as a random range for a sample of size $n$ which is taken from a standard normal population $N(0,1)$. If we divide $E[R_n]$ and $\sigma_{R_n}$ by $\sigma$, we will obtain the mean range and the standard deviation of range, respectively, for $N(0,1)$. Therefore, to compute $P_0$ we can use the standard normal distribution. Table 2.11 gives the $P_0$ values for sample sizes $n = 4 - 10$. For example, for $n = 4$, $P_0 \approx 0.0049$ and ARL $\approx 205$.

Denote

$$\gamma = \frac{\sigma_m}{\sigma_{pr}}. \tag{2.8.9}$$

It follows from (2.8.7) that now the measurement $Y \sim (\mu, \sigma_{pr}^2(1 + \gamma^2))$.

Suppose that $\gamma = 0.37$. Let us find out the corresponding ARL for $n = 4$. The control limit equals $4.699 \times \sigma_{R_n}$ (see Table 2.11, column 5). This limit was established for the ideal situation where $\sigma_m = 0$, or $\gamma = 0$. Now the sample range of random variable $Y$ has increased by a factor $\sqrt{1 + \gamma^2}$.

It can be shown that now the probability of crossing the control limit equals

$$P_\gamma = P\big(R_n > 4.699/\sqrt{1+\gamma^2}\big). \tag{2.8.10}$$

For our example, $\sqrt{1+\gamma^2} = \sqrt{1+0.37^2} = 1.066$, and we have to find $P(R_n > 4.699/1.066) \approx P(R_n > 4.41)$. This probability can be found using special tables; see Dunin-Barkovsky and Smirnov (1956, p. 514). The desired quantity is $\approx 0.01$, from which it follows that ARL $\approx 100$. Thus, if the standard deviation of the measurement instrument is about 37% of the process standard deviation, the ARL of the $R$ chart decreases by a factor of approximately 0.5.

### Performance of $X$ Bar Chart in the Presence of Measurement Bias

Let us investigate the influence of measurement bias $\delta_{instr}$ and variance $\sigma_m^2$ on the probability of crossing the control limits LCL $= \mu - 3\sigma_{pr}/\sqrt{n}$ and UCL $= \mu + 3\sigma_{pr}/\sqrt{n}$.

This material, like the material at the end of the previous subsection, is of theoretical nature and can be omitted when first reading the book. Only the final numerical illustrations are important.

Suppose that in (2.8.4) the process shift $\delta_{pr} = 0$, and the measurement instrument introduces a systematic error $\delta_{instr}$. Then $Y \sim N(\mu + \delta_{instr}, \sigma_{pr}^2 + \sigma_m^2)$. The average $\overline{Y}$ of $n$ observations is then a normally distributed random variable with mean $\mu + \delta_{instr}$ and standard deviation $\sigma_{\overline{Y}} = \sigma_{pr}\sqrt{1+\gamma^2}/\sqrt{n}$. Therefore

$$Z_0 = (\overline{Y} - \mu - \delta_{instr})/\sigma_{\overline{Y}} \sim N(0,1). \tag{2.8.11}$$

Now the probability that the sample average is between LCL and UCL equals

$$P\left(\text{LCL} < \overline{Y} < \text{UCL}\right) \tag{2.8.12}$$

$$= P\left(\frac{-3\sigma_{pr} - \delta_{instr}\sqrt{n}}{\sigma_{pr}\sqrt{1+\gamma^2}} < Z_0 < \frac{3\sigma_{pr} - \delta_{instr}\sqrt{n}}{\sigma_{pr}\sqrt{1+\gamma^2}}\right).$$

After some simple algebra we obtain that the probability that $\overline{Y}$ is outside the control limits equals

$$P\big(\overline{Y} < \text{LCL or } \overline{Y} > \text{UCL}\big) = \Phi\left(\frac{\beta\sqrt{n}-3}{\sqrt{1+\gamma^2}}\right) + \Phi\left(\frac{-\beta\sqrt{n}-3}{\sqrt{1+\gamma^2}}\right), \tag{2.8.13}$$

where $\beta = \delta_{instr}/\sigma_{pr}$ and $\Phi(\cdot)$ is the distribution function for the standard normal variable; see (2.3.3).

*Example 2.8.1: The ARL for the $X$ bar chart in the presence of instrument bias*
Assume that the sample size is $n = 4$, $\gamma = \sigma_m/\sigma_{pr} = 0.2$ and $\beta = 0.5$. Thus we assume that the measurement bias is half of $\sigma_{pr}$, and the standard deviation of measurement error is 20% of the process standard deviation.

Substituting these values into (2.8.13) we obtain, using the table in Appendix A and assuming $\Phi(3.92) = 1.000$, that

$$P\left(\overline{Y} < \text{LCL or } \overline{Y} > \text{UCL}\right) = \Phi(-1.96) - \Phi(-3.92) = 1 - 0.975 = 0.025.$$

The ARL is therefore $1/0.025 = 40$. Let us compare this with the ARL for an "ideal" measurement instrument ($\delta_{instr} = \sigma_m = 0$).

The probability that the averages go outside three-sigma limits equals $2\Phi(-3) = 2(1 - 0.9986) = 0.0028$. The corresponding ARL $= 1/0.0028 \approx 360$. We see therefore that the presence of $\delta_{instr} = 0.5\sigma_{pr}$ and $\sigma_m = 0.2\sigma_{pr}$ reduces the ARL by a factor of 9.

## 2.9  Exercises

**1.** Below are the results of 15 replicate determinations of nitrate ion concentration in mg/ml in a particular water specimen:

$$0.64, 0.64, 0.63, 0.63, 0.62, 0.65, 0.66, 0.63, 0.60, 0.64, 0.65, 0.66, 0.61, 0.62, 0.63.$$

    **a.** Find estimate $\hat{\mu}$ of the mean concentration.

    **b.** Estimate the standard deviation via formula (2.1.7) and via the sample range. For the latter divide the sample range by $A_{15} = 3.472$.

    **c.** Would you consider the smallest result as an outlier (use Dixon's test)?

    **d.** Construct a 95-% confidence interval on the mean concentration.

    **e.** Using the rule of Sect. 2.2, check whether the original data are recorded with sufficient accuracy.

**2.** It is known that the weight of Yarok apples is normally distributed with mean 150 g and standard deviation 15 g. Find the probability that the weight of a given apple will be inside the interval $[145, 158]$.

**3.** In order to avoid systematic error in the process of weighing, the following method is used in analytic chemistry. First, a vessel containing a liquid is weighed, and then the weight of the empty vessel is subtracted from the previous result. Suppose that the random weighing error has a standard deviation $\sigma = 0.0005$ g. What will be the standard deviation of the weight difference?
*Solution:* Using (2.1.22), with $a_1 = 1, a_2 = -1$, obtain that the standard deviation of the weight difference is $0.0005 \times \sqrt{2} = 0.0071$.

**4.** Stigler (1977) presents Simon Newcomb's measurements of the speed of light carried out in 1882. The data are presented as deviations in nanoseconds from 24 800, the time for the light to travel 7442 meters. Table 2.12 gives the first 30 measurements (out of 66):[1]

---

[1] Reprinted with permission from Andrew Gelman, John B. Carlin, Hal S. Stern and Donald Rubin, *Bayesian Data Analysis* (2000), p. 70, Copyright CRC Press, Boca Raton, Florida

Table 2.12: Newcomb's light speed data (Stigler 1977)

| 28 | 26 | 33 | 24 | 34 | −44 | 27 | 16 | 40 | −2 |
|----|----|----|----|----|-----|----|----|----|----|
| 29 | 22 | 24 | 21 | 25 |  30 | 23 | 29 | 31 | 19 |
| 24 | 20 | 36 | 32 | 36 |  28 | 25 | 21 | 28 | 29 |

Assume the normal model and check whether the lowest measurement −44 might be considered as an outlier.

*Solution.* The Dixon's statistic $Q = (-2 - (-44)/(36 - (-44)) = 0.575 > C_{30} = 0.260$; see Table 2.5.

**5.** In order to check the titration procedure for a potential bias in measuring NaCl concentration in water, a standard solution with 0.5% concentration of NaCl was tested 7 times, and the following percentage results were obtained:

0.51, 0.55, 0.53, 0.50, 0.56, 0.53, 0.57.

Test the null hypothesis that the concentration is 0.5% at the significance level 0.05. Use for this purpose the 0.95- confidence interval for the mean concentration. Estimate the measurement bias.

*Solution.* The average concentration is 0.536, $s = 0.026$. The 95% confidence interval [0.512, 0.560] does not contain the point 0.5. The estimate of the bias is $\hat{\delta} = 0.536 - 0.500 = 0.036 \approx 0.04$.

# Chapter 3

# Comparing Means and Variances

*Our discontent is from comparison.*

J. Norris

## 3.1  $t$-test for Comparing Two Population Means

In this section we consider a very important and widely used test in statistics called the $t$-test which is designed to compare the mean values in two independent normally distributed populations A and B. Let us first consider an example.

*Example 3.1.1: Dextroamphetamine excretion by children*[1]
This example (Devore, 1982, p. 292) gives data on the amount of a special chemical substance called dextroamphetamine (DEM) excreted by children. It is assumed that the children with organically related disorders produce more DEM than the children with nonorganic disorders. To test this assumption, two samples of children were chosen and both were given a drug containing DEM. These samples were compared by the percentage of drug recovery seven hours after its administration.

---

[1] Borrowed from *Probability and Statistics for Engineering and the Sciences*, 1st Edition by J. Devore. ©1982. Reprinted with permission of Brooks/Cole, a division of Thompson Learning: www.thompsonsrights.com. Fax 800 730-2215.

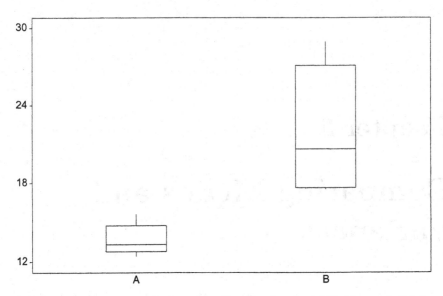

Figure 3.1. Box and whisker plots for samples A and B, Example 3.1.1

Sample A (nonorganic disorders): $15.59, 14.76, 13.32, 12.45, 12.79$.

Sample B (organic disorders): $17.53, 20.60, 17.62, 28.93, 27.10$.

Figure 3.1 shows the box and whisker plot for both samples. We see that the samples differ considerably in their mean values and in their standard deviations. We assume therefore that populations A and B are distributed as $N(\mu_A, \sigma_A^2)$ and $N(\mu_B, \sigma_B^2)$, respectively, with $\sigma_A \neq \sigma_B$.

Let us define the null hypothesis $\mathcal{H}_0$: $\mu_A = \mu_B$ and the alternatives

$\mathcal{H}_- : \mu_A < \mu_B$,

$\mathcal{H}_+ : \mu_A > \mu_B$ and

$\mathcal{H}^\star : \mu_A \neq \mu_B$.

The test is based on the following statistic:

$$T = \frac{\overline{x}_A - \overline{x}_B}{\sqrt{s_A^2/n_A + s_B^2/n_B}}, \tag{3.1.1}$$

where $\overline{x}_A$ and $\overline{x}_B$ are the sample means, and $s_A^2$ and $s_B^2$ are sample variances, computed by formulas (2.1.6) and (2.1.8), respectively. $n_A$ and $n_B$ are the sample sizes.

Define also the number of degrees of freedom $\nu$:

$$\nu = \frac{(s_A^2/n_A + s_B^2/n_B)^2}{(s_A^2/n_A)^2/(n_A - 1) + (s_B^2/n_B)^2/(n_B - 1)}. \tag{3.1.2}$$

The value of $\nu$ must be rounded to the nearest integer.

Table 3.1: Critical values of $t_\alpha(\nu)$ of the $t$-distribution

| $\nu$ | $\alpha = 0.05$ | $\alpha = 0.025$ | $\alpha = 0.01$ | $\alpha = 0.005$ |
|---|---|---|---|---|
| 1 | 6.314 | 12.706 | 31.821 | 63.657 |
| 2 | 2.920 | 4.303 | 6.965 | 9.925 |
| 3 | 2.353 | 3.182 | 4.451 | 5.841 |
| 4 | 2.132 | 2.776 | 3.747 | 4.604 |
| 5 | 2.015 | 2.571 | 3.365 | 4.032 |
| 6 | 1.943 | 2.447 | 3.143 | 3.707 |
| 7 | 1.895 | 2.365 | 2.998 | 3.499 |
| 8 | 1.860 | 2.306 | 2.896 | 3.355 |
| 9 | 1.833 | 2.262 | 2.821 | 3.250 |
| 10 | 1.812 | 2.228 | 2.764 | 3.169 |
| 11 | 1.796 | 2.201 | 2.718 | 3.106 |
| 12 | 1.782 | 2.179 | 2.681 | 3.055 |
| 13 | 1.771 | 2.160 | 2.650 | 3.012 |
| 14 | 1.761 | 2.145 | 2.624 | 2.977 |
| 15 | 1.753 | 2.131 | 2.602 | 2.947 |
| 16 | 1.746 | 2.120 | 2.583 | 2.921 |
| 17 | 1.740 | 2.110 | 2.567 | 2.898 |
| 18 | 1.734 | 2.101 | 2.552 | 2.878 |
| 19 | 1.729 | 2.093 | 2.539 | 2.861 |
| 20 | 1.725 | 2.086 | 2.528 | 2.845 |
| 25 | 1.708 | 2.060 | 2.485 | 2.787 |
| 30 | 1.697 | 2.042 | 2.457 | 2.750 |
| 40 | 1.697 | 2.021 | 2.423 | 2.704 |

Denote by $t_\alpha(\nu)$ the $\alpha$-critical value of the $t$-distribution with $\nu$ degrees of freedom. We assume that the reader is familiar with this distribution, often referred to as Student's distribution. If a random variable $T_0$ has a $t$-distribution with $\nu$ degrees of freedom, the probability that $T_0$ exceeds $t_\alpha(\nu)$ is equal to $\alpha$. The $t$-distribution is symmetric with respect to zero, and therefore the probability that $T_0$ is less than $-t_\alpha(\nu)$ also equals $\alpha$. Table 3.1 gives the values of $t_\alpha(\nu)$. We reject $\mathcal{H}_0$ in favor of $\mathcal{H}_-$ at the significance level $\alpha$ if the value of $T$ computed by (3.1.1) is less than or equal to $-t_\alpha(\nu)$. We reject $\mathcal{H}_0$ in favor of $\mathcal{H}_+$ at level $\alpha$ if the value of $T$ is greater than or equal to $t_\alpha(\nu)$. We reject $\mathcal{H}_0$ in favor of $\mathcal{H}^\star$ at level $\alpha$ if the absolute value of $T$ is greater than or equal to $t_{\alpha/2}(\nu)$.

*Example 3.1.1 continued*
Let us complete the calculations for Example 3.1.1. We have $\bar{x}_A = 13.782$, $s_A = 1.34$, $n_A = 5$. Similarly, $\bar{x}_B = 22.356$ $s_B = 5.35$, $n_B = 5$. (Observe that the

populations differ quite significantly in the values of their estimated standard deviations.) By (3.1.1) the test statistic is

$$T = \frac{13.782 - 22.356}{\sqrt{1.34^2/5 + 5.35^2/5}} = -3.477$$

Now let us compute $\nu$. By (3.1.2),

$$\nu = \frac{(1.34^2/5 + 5.35^2/5)^2}{1.34^4/(5^2 \times 4) + 5.35^4/(5^2 \times 4)} = 4.5 \approx 5.$$

We will check $\mathcal{H}_0$ against $\mathcal{H}_- : \mu_A < \mu_B$ at significance $\alpha = 0.01$. We see from Table 3.1 that $T = -3.477$ is *less* than $-t_{0.01}(5) = -3.365$ and thus we reject the null hypothesis. We confirm therefore the assumption that DEM excretion for nonorganic disorders is smaller than for organic disorders.

### Remark 1
What if the normality assumption regarding both populations is not valid? It turns out that the $t$-test used in the above analysis is not very sensitive to deviations from normality. More on this issue can be found in Box (1953), and an enlightening discussion can be found in Scheffé (1956, Chapter 10). It is desirable to use, in addition to the $t$-test, the so-called *nonparametric* approach, such as a test based on ranks. We discuss in Sect. 3.2 a test of this kind which is used to compare mean values in two or more populations.

### Remark 2
It is quite typical of chemical measurements for there to be a *bias* or a systematic error caused by a specific operation method in a specific laboratory. The bias may appear as a result of equipment adjustment (e.g. a shift of the zero point), operational habits of the operator, properties of reagents used, etc. It is vitally important for the proper comparison of means to ensure that the two sets of measurement (sample A and sample B) have *the same* systematic errors. Then, in the formula for the $T$-statistic, the systematic errors in both samples cancel out. Interestingly, this comment is rarely made in describing the applications of the $t$-test.

In order to ensure the equality of systematic errors, the processing of samples A and B must be carried out on the same equipment, in the same laboratory, possible by the same operator, and within a relatively short period of time. These conditions are referred to in measurement practice as "repeatability conditions"; see clause 4.2.1 "Two groups of measurements in one laboratory", of the British Standard BS ISO 5725-6 (1994).

### Remark 3
Is there a way to estimate bias? In principle, yes. Prepare a reference sample with a known quantity of the chemical substance. Divide this sample into $n$ portions, and carry out, in repeatability conditions, the measurements for each

of these $n$ portions. Calculate the mean quantity $\bar{x}_n$. If $x_0$ is the known contents of the chemical substance, then the difference $\hat{\Delta} = x_0 - \bar{x}_n$ is an estimate of measurement bias. Construct, for example, the confidence interval for the mean difference $E[\Delta] = \mu$ and check whether it contains the zero point. If not, assume that the bias does exist and $\hat{\Delta}$ is its estimate.

Suppose that we can assume that the variances in populations A and B are equal: $\sigma_A = \sigma_B$. Then the testing procedure for hypotheses about $\mu_A - \mu_B$ is similar to that described above, with some minor changes. The test statistic will be

$$T = \frac{\bar{x}_A - \bar{x}_B}{s_p \sqrt{1/n_A + 1/n_B}}, \tag{3.1.3}$$

where $s_p$ (called the pooled sample variance) is defined as

$$s_p = \frac{(n_A - 1)s_A^2 + (n_B - 1)s_B^2}{n_A + n_B - 2}, \tag{3.1.4}$$

in which $s_A^2$ and $s_B^2$ are the sample variances for samples A and B, respectively, calculated as in (2.1.8). The number of degrees of freedom for this version of the $t$-test is $\nu = n_A + n_B - 2$.

## Paired Experiment for Comparing Means

Suppose that we wish to compare two wear-resistant materials, A and B. For this purpose, we prepare two samples of 8 specimens from the above materials. Suppose that testing these specimens is done by two laboratories, 1 and 2. Suppose that each of the laboratories tests 8 specimens. One way of organizing the experiment is to give sample A to lab 1 and the sample B to lab 2. Suppose that the results differ significantly, and that sample A is definitely better, i.e. shows less wear. Would the results of this experiment be conclusive in favor of material A, if the corresponding $t$-test indicates that the wear in sample A is significantly smaller than in sample B? On one hand yes, but on the other there remain some doubts that possibly lab 1 applied smaller friction force than lab 2, or used different abrasive materials, etc. So, it is desirable to organize the experiment in such a way that the possible differences in the testing procedure are eliminated.

An efficient way to achieve this goal is to redesign the experiment in the following way. Choose randomly 8 pairs of specimens, each pair containing one from sample A and one from sample B. Organize the experiment in such a way that each pair is tested in identical conditions. Compute for each pair *the difference* in the amount of wear between the two specimens. In this way, the differences resulting from the testing conditions will be eliminated. What will be measured is the "pure" difference in the behaviour of the materials. This type of experiment is called "pairwise blocking".

Table 3.2: Experiment 1: Average weight loss in % for material A

| Driver | 1 | 2 | 3 | 4 | 5 | 6 | 7 | 8 |
|---|---|---|---|---|---|---|---|---|
| Weight loss | 5.0 | 7.5 | 8.0 | 6.5 | 10.0 | 9.5 | 12.5 | 11.0 |

Table 3.3: Experiment 2: Average weight loss in % for material B

| Driver | 1 | 2 | 3 | 4 | 5 | 6 | 7 | 8 |
|---|---|---|---|---|---|---|---|---|
| Weight loss | 5.5 | 8.5 | 9.0 | 7.0 | 11.0 | 9.5 | 13.5 | 11.5 |

To make the advantages of pairwise blocking more obvious, let us consider an imaginary experiment which is a variation on the famous statistical experiment for comparing the wear of shoe soles made from two different materials.

*Example 3.1.2: Comparing the wear of two materials for disk brake shoes*
First, consider a "traditional" two-sample approach. Two groups of $n = 8$ similar cars (same model, same production year) are chosen. The first (group A) has front wheel brake shoes made of material A (each car has two pairs of shoes, one pair for each front wheel). Each particular shoe is marked and weighed before its installation in the car. Eight drivers are chosen randomly from the pool of drivers available, and each driver is assigned to a certain car.

In the first experiment, each driver drives his car for 1000 miles. Afterwards the brake shoes are removed, their weight is compared to the initial weight, the relative loss of weight for each shoe is calculated, and the average loss of weight for all four shoes for each car is recorded. In the second experiment, the same cars (each with the same driver) are equipped with brake shoes made of material B, and the whole procedure is repeated. To exclude any driver–material interaction, the drivers do not know which material is used in each experiment.

The results obtained for Experiments 1 and 2 are presented in Tables 3.2 and 3.3.

It is obvious from Fig. 3.2 that material B shows greater wear than material A. Moreover, this claim is not only true "on average", but also holds for each car.

The amount of wear has a large variability caused by shoe material nonhomogeneity as well as by the variations in driving habits of different drivers and differences in the roads driven by the cars. From the statistical point of view, the crucial fact is that this variability *hides* the true difference in the wear.

To show this, let us apply the above-described $t$-test to test the null hypothesis $\mu_A = \mu_B$ against the obvious alternative $\mu_A < \mu_B$. (Here index A refers to Experiment 1 and index B to experiment 2.)

Table 3.4: The difference of wear measured by average weight loss in percentage

| Driver | 1 | 2 | 3 | 4 | 5 | 6 | 7 | 8 |
|--------|-----|-----|-----|-----|-----|-----|-----|-----|
| Weight loss | 0.4 | 1.1 | 1.6 | 0.6 | 0.9 | 0.1 | 1.2 | 0.4 |

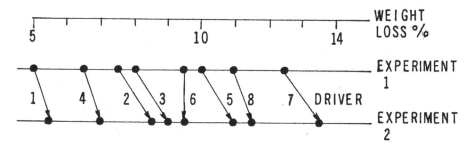

Figure 3.2. Comparison of wear for Experiment 1 and 2

We calculate that $x_A = 8.75$, $x_B = 9.44$, $s_A^2 = 6.07$ and $s_B^2 = 6.53$. We have $n_A = n_B = 8$, $\nu = 14$. By (3.11) we obtain

$$T = \frac{8.75 - 9.44}{\sqrt{6.07/8 + 6.53^2/8}} = -0.55.$$

The 0.05-critical value $t_{0.05}(14) = 1.761$. We must reject the null hypothesis only if the computed value of $T$ is *smaller* than $-1.761$. This is not the case, and therefore our $t$-test fails to confirm that material B shows greater wear than material A.

Now let us consider a more clever design for the whole experiment. The disk brake has two shoes. Let us make one shoe from material A and the other from material B. Suppose that 8 cars take part in the experiment, and that the same drivers take part with their cars. Each car wheel is equipped with shoes made of different materials (A and B), and the location (outer or inner) of the A shoe is chosen randomly.

For each car, we record the *difference* in the average weight loss for shoes of material A and B, i.e. we record the average weight loss of B shoes minus the average weight loss of A shoes. Table 3.4 shows simulated results for this imaginary experiment. For each car, the data in this table were obtained as the difference between previously observed values perturbed by adding random numbers.

There is no doubt that these data point out on a significant advantage of material A over material B. The calculations show that the average difference in wear is $\overline{\Delta} = 0.79$ and the sample variance $s_\Delta^2 = 0.25$. Let us construct a 95% confidence interval for the mean weight loss difference $\mu_\Delta$; see Sect. 2.7. The confidence interval has the form

$$(\overline{\Delta} - t_{0.025}(7) \cdot s_\Delta/\sqrt{8}, \overline{\Delta} + t_{0.025}(7) \cdot s_\Delta/\sqrt{8})$$
$$= (0.79 - 2.365 \cdot 0.5/2.83, 0.79 + 2.365 \cdot 0.5/2.83) = (0.37, 1.21).$$

We obtain therefore that the confidence interval *does not* contain zero, and the null hypothesis must certainly be rejected.

## 3.2   Comparing More Than Two Means

In this section we consider the comparison of several mean values by means of a simple yet efficient procedure called the Kruskal–Wallis (KW) test. Unlike the tests considered so far, the KW test is based not on the actual observations but on their *ranks*. It belongs to the so-called nonparametric methods.

Let us remind the reader what we mean by ranks. Suppose that we have a sample of 8 values: $\{12, 9, 14, 17, 7, 25, 5, 18\}$. Now put these values in increasing order:

$(5, 7, 9, 12, 14, 17, 18, 25)$

and number them, from left to right. The rank of any observation will be its ordinal number. So, the rank of 5, $r(5)$, is 1, the rank of 7 is $r(7) = 2$, etc.

How to assign the ranks if several observations are tied, i.e. are equal to each other, as in the following sample: $\{12, 12, 14, 17, 7, 25, 5, 18\}$?

Order the sample, assign an ordinal number to each observation, and define the rank of a tied observation as the corresponding average rank:

| Ordered sample | 5, | 7, | 12, | 12, | 14, | 17, | 18, | 25; |
|---|---|---|---|---|---|---|---|---|
| Ordinal numbers | 1, | 2, | 3, | 4, | 5, | 6, | 7, | 8; |
| Ranks | 1, | 2, | 3.5, | 3.5, | 5, | 6, | 7, | 8. |

### Notation for the KW Test

We consider $I$ samples, numbered $i = 1, 2, \ldots, I$; sample $i$ has $J_i$ observations. $N = J_1 + \ldots + J_I$ is the total number of observations; $x_{ij}$ is the $j$th observation in the $i$th sample.

It is assumed that the $i$th sample comes from the population described by random variable

$$X_i = \mu_i + \epsilon_{ij}, \, i = 1, \ldots, I, \, j = 1, 2, \ldots, J_i, \qquad (3.2.1)$$

where all random variables $\epsilon_{ij}$ have *the same* continuous distribution. Without loss of generality, it can be assumed that $\mu_i$ is the mean value of $X_i$.

It is important to stress that the Kruskal–Wallis procedure does not demand that the random samples be drawn from normal populations. It suffices to demand that the populations involved have the same distribution but may differ in their location.

Suppose that all $x_{ij}$ values are pooled together and ranked in increasing order. Denote by $R_{ij}$ the *rank* of the observation $x_{ij}$. Denote by $R_i$. the total rank (the sum of the ranks) of all observations belonging to sample $i$. $\overline{R}_{i\cdot} = R_i./J_i$ is the average rank of sample $i$.

The null hypothesis $\mathcal{H}_0$ is that all $\mu_i$ are equal:

$$\mu_1 = \mu_2 = \ldots = \mu_I. \tag{3.2.2}$$

In the view of the assumption that all $\epsilon_{ij}$ have the same distribution, the null hypothesis means that all $I$ samples belong to the same population.

If $\mathcal{H}_0$ is true, one would expect the values of $\overline{R}_i$ to be close to each other and hence close to the overall average

$$R_{..} = \frac{\overline{R}_{1\cdot} + \ldots + \overline{R}_{I\cdot}}{N} = \frac{N+1}{2}. \tag{3.2.3}$$

An appropriate criterion for measuring the overall closeness of sample rank averages to $R_{..}$ is a weighted sum of squared differences $(\overline{R}_{i\cdot} - R_{..})^2$.

The Kruskal–Wallis statistic is given by the following expression:

$$KW = \frac{12}{N(N+1)} \sum_{i=1}^{I} J_i \left(\overline{R}_{i\cdot} - \frac{N+1}{2}\right)^2. \tag{3.2.4}$$

The use of the KW test is based on the following:

**Proposition 3.2.1**

When the null hypothesis is true and either $I = 3, J_i \geq 6$, $i = 1, 2, 3$, or $I > 3, J_i \geq 5$, $i = 1, 2, \ldots, I$, then $KW$ has approximately a chi-square distribution with $\nu = I - 1$ degrees of freedom (see Devore 1982, p. 597).

The alternative hypothesis for which the KW test is most powerful is the following:

$\mathcal{H}^*$: not all population means $m_1, \ldots, m_I$ are equal.

Since $KW$ is zero when all $R_i$. are equal and is large when the samples are shifted with respect to each other, the null hypothesis is rejected for large values of $KW$. According to the above Proposition 3.2.1, the null hypothesis is rejected at significance level $\alpha$ if $KW > \chi^2_\alpha(\nu)$, where $\chi^2_\alpha(\nu)$ is the $1 - \alpha$ quantile of the chi-square distribution with $\nu$ degrees of freedom.

*Remark 1: Chi-square distribution*
Let us remind the reader that the chi-square distribution is defined as the distribution of the sum of squares of several standard normal random variables.

We say that the random variable $G$ has a chi-square distribution with $k$ degrees of freedom if $G = X_1^2 + \ldots + X_k^2$, where all $X_i$ are independent random

Table 3.5: The values of $\chi^2_\alpha(\nu)$

| d.f. $\nu$ | $\alpha = 0.1$ | $\alpha = 0.05$ | $\alpha = 0.025$ | $\alpha = 0.01$ | $\alpha = 0.005$ |
|---|---|---|---|---|---|
| 1 | 2.706 | 3.841 | 5.024 | 6.635 | 7.879 |
| 2 | 4.605 | 5.991 | 7.378 | 9.210 | 10.597 |
| 3 | 6.251 | 7.815 | 9.348 | 11.345 | 12.838 |
| 4 | 7.779 | 9.488 | 11.143 | 13.277 | 14.860 |
| 5 | 9.236 | 11.070 | 12.833 | 15.086 | 16.750 |
| 6 | 10.645 | 12.592 | 14.449 | 16.812 | 18.548 |
| 7 | 12.017 | 14.067 | 16.013 | 18.475 | 20.278 |
| 8 | 13.362 | 15.507 | 17.535 | 20.090 | 21.955 |
| 9 | 14.684 | 16.919 | 19.023 | 21.666 | 23.589 |
| 10 | 15.987 | 18.307 | 20.483 | 23.209 | 25.188 |
| 11 | 17.275 | 19.675 | 21.920 | 24.725 | 26.757 |
| 12 | 18.549 | 21.026 | 23.337 | 26.217 | 28.300 |
| 13 | 19.812 | 22.362 | 24.736 | 27.688 | 28.819 |
| 14 | 21.064 | 23.685 | 26.119 | 29.141 | 31.319 |
| 15 | 22.307 | 24.996 | 27.488 | 30.578 | 32.801 |

variables distributed as $N(0,1)$. The corresponding critical values are defined as follows:

$$\alpha = P\big(G > \chi^2_\alpha(k)\big). \tag{3.2.5}$$

Table 3.5 gives the critical values of $\chi^2_\alpha(\nu)$ for the KW test. A more complete table of the quantiles of the chi-square distribution is presented in Appendix B. Let us consider an example.

*Example 3.2.1: Silicon wafer planarity measurements* [1]

Silicon wafers undergo a special chemical-mechanical planarization procedure in order to achieve ultra-flat wafer surfaces. To control the process performance, a sample of wafers from one batch was measured at nine sites. Table 3.6 presents a fragment of a large data set, for five wafers and for four sites.

Assuming that the wafer thickness at different sites can differ only by a shift parameter, let us check the null hypothesis $\mathcal{H}_0$ that the thickness has the same mean value at all four sites.

The mean ranks are 4.6, 8.0, 13.8 and 15.6 for sites 1 through 4, respectively. The KW statistic equals 11.137 on four degrees of freedom. From Table 3.5, it

---

[1] Source: Arnon M. Hurwitz and Patrick D. Spagon "Identifying sources of variation", pp. 105-114, in the collection by Veronica Czitrom and Patrick D. Spagon *Statistical Case Studies for Industrial Process Improvement* ©1997. Borrowed with the kind permission of the ASA and SIAM

Table 3.6: Thickness (in angstroms) for four sites on the wafer

| Wafer | Site 1 | Site 2 | Site 3 | Site 4 |
|-------|--------|--------|--------|--------|
| 1 | 3238.8 | 3092.7 | 3320.1 | 3487.8 |
| 2 | 3201.4 | 3212.6 | 3365.6 | 3291.3 |
| 3 | 3223.7 | 3320.8 | 3406.2 | 3336.8 |
| 4 | 3213.1 | 3300.0 | 3281.6 | 3312.5 |
| 5 | 3123.6 | 3277.1 | 3289.1 | 3369.6 |

corresponds to $\alpha \approx 0.025$. We reject the null hypothesis at $\alpha = 0.05$.

How to proceed if the null hypothesis is rejected? The statistician's first task is to find those samples which do not show significant dissimilarities with respect to their mean values. For this purpose a statistical procedure called "multiple comparisons" is used (see Devore 1982, p. 598). Multiple comparisons are based on a pairwise comparisons between all pairs of $I$ samples involved. To avoid tedious calculations, the use of statistical software is recommended. Every statistical package has an option for multiple comparisons in the Kruskal–Wallis procedure. Using Statistix reveals that samples 1, 2 and 3, or alternatively samples 2, 3, 4 may be considered, at significance level $\alpha = 0.05$, as having the same mean values. The box and whisker plot in Fig. 3.3 also suggests that all four samples cannot be treated as coming from the same population.

The next step in data analysis would be an investigation of the production process. Finding and eliminating the factors responsible for large dissimilarities in the wafer thickness at different sites would be a joint undertaking by engineers and statisticians.

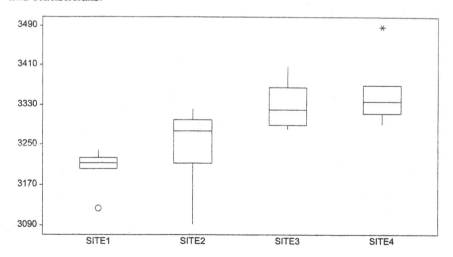

Figure 3.3. Box and whisker plot of wafer thickness at sites 1,2,3 and 4.

Table 3.7: Critical values of Hartley's statistic, for $\alpha = 0.05$ (Sachs 1972, Table 151)

| $n-1$ | $k=2$ | $k=3$ | $k=4$ | $k=5$ | $k=6$ | $k=7$ | $k=8$ | $k=9$ |
|---|---|---|---|---|---|---|---|---|
| 2 | 39.0 | 87.5 | 142 | 202 | 266 | 333 | 403 | 475 |
| 3 | 15.40 | 27.8 | 39.2 | 50.7 | 62.0 | 72.9 | 83.5 | 93.9 |
| 4 | 9.60 | 15.5 | 20.6 | 25.2 | 29.5 | 33.6 | 37.5 | 41.1 |
| 5 | 7.15 | 10.8 | 13.7 | 16.3 | 18.7 | 20.8 | 22.9 | 24.7 |
| 6 | 5.82 | 8.38 | 10.4 | 12.1 | 13.7 | 15.0 | 16.3 | 17.5 |
| 7 | 4.99 | 6.94 | 8.44 | 9.70 | 10.8 | 11.8 | 12.7 | 13.5 |
| 8 | 4.43 | 5.34 | 6.31 | 7.11 | 7.80 | 8.41 | 8.95 | 11.1 |
| 9 | 4.03 | 5.34 | 6.31 | 7.11 | 7.80 | 8.41 | 8.95 | 9.45 |
| 10 | 3.72 | 4.85 | 5.67 | 6.34 | 6.92 | 7.42 | 7.87 | 8.28 |
| 12 | 3.28 | 4.16 | 4.79 | 5.30 | 5.72 | 6.09 | 6.42 | 6.72 |

## 3.3 Comparing Variances

We present in this section two popular tests for the hypothesis of equality of several variances.

### Hartley's Test

Assume that we have $k$ samples of *equal size* $n$ drawn from independent normal populations. Let $s_1^2, s_2^2, \ldots, s_k^2$ be the estimates of the corresponding variances computed using (2.1.8).

To test the null hypothesis $\mathcal{H}_0 : \sigma_1^2 = \sigma_2^2 = \ldots = \sigma_k^2$, we have to compute the following statistic due to Hartley:

$$\hat{F}_{max} = \frac{\max s_i^2}{\min s_i^2}. \tag{3.3.1}$$

The critical values of the $\hat{F}_{max}$ statistic are given in Table 3.7. The null hypothesis is rejected (at the significance level $\alpha = 0.05$) in favor of an alternative that at least one of variances is different from others, if the observed value of the Hartley's statistic *exceeds* the corresponding critical value in Table 3.7.

*Example 3.3.1: Testing the equality of variances for wafer thickness data*
Using formula (2.1.8), we compute the variances for the thickness of sites 1 through 4. In this example, the number of samples $k = 4$, and the sample size $n = 5$. We obtain the following results:

$$s_1^2 = 2019, \quad s_2^2 = 8488, \quad s_3^2 = 2789, \quad s_4^2 = 5985.$$

Hartley's statistic equals $\hat{F}_{max} = 8488/2019 = 4.2$. The corresponding critical value for $n - 1 = 4$ and $k = 4$ is 20.6. Since this number exceeds the observed value of the statistic, the null hypothesis is *not* rejected.

### Bartlett's Test

A widely used procedure is Bartlett's test. Unlike Hartley's test, it is applicable for samples of unequal size. The procedure involves computing a statistic whose distribution is closely approximated by the chi-square distribution with $k - 1$ degrees of freedom, where $k$ is the number of random samples from independent normal populations.

Let $n_1, n_2, \ldots, n_k$ be the sample sizes, and $N = n_1 + \ldots + n_k$. The test statistic is

$$B = (\log_e 10) \cdot \frac{q}{c}, \tag{3.3.2}$$

where

$$q = (N - k) \log_{10} s_p^2 - \sum_{i=1}^{k} (n_i - 1) \log_{10} s_i^2, \tag{3.3.3}$$

$$c = 1 + \frac{1}{3(k - 1)} \left( \sum_{i=1}^{k} (n_i - 1)^{-1} - (N - k)^{-1} \right) \tag{3.3.4}$$

and $s_p^2$ is the so-called pooled variance of all sample variances $s_i^2$, given by

$$s_p^2 = \frac{\sum_{i=1}^{k} (n_i - 1) s_i^2}{N - k}. \tag{3.3.5}$$

The quantity $q$ is large when the sample variances $s_i^2$ differ greatly and is equal to zero when all $s_i^2$ are equal. We reject $\mathcal{H}_0$ at the significance level $\alpha$ if

$$B > \chi_\alpha^2(k - 1). \tag{3.3.6}$$

Here $\chi_\alpha^2(k - 1)$ is the $1 - \alpha$ quantile of the chi-square distribution with $k - 1$ degrees of freedom. These values are given in Table 3.5.

It should be noted that Bartlett's test is very sensitive to deviations from normality and should not be applied if the normality assumption of the populations involved is doubtful.

*Example 3.3.2 continued*
By (3.3.5), $s_p^2 = 4820.2$. By (3.3.3), $q = 1.104$, and by (3.3.4) $c = 1.104$. Thus the test statistic

$$B = \log_e 10 \cdot \tfrac{1.104}{1.104} = 2.303.$$

From Table 3.5 we see that the critical value of chi-square for $k - 1 = 4$ and $\alpha = 0.05$ is 9.49. Thus the null hypothesis is not rejected.

Table 3.8: Diastolic blood pressure measurement results

| person no | Left hand | Right hand |
|-----------|-----------|------------|
| 1 | 72 | 74 |
| 2 | 75 | 75 |
| 3 | 71 | 72 |
| 4 | 77 | 78 |
| 5 | 78 | 76 |
| 6 | 80 | 82 |
| 7 | 76 | 76 |
| 8 | 75 | 78 |

## 3.4  Exercises

**1.** Consider the first and and the fourth site data in Table 3.6. Use the $t$-test to check the hypothesis that $\mu_1 = \mu_4$.

**2.** Use Hartley's test to check the null hypothesis on the equality of variances in five populations of day 1 through day 5; see Table 4.1.

**3.** It is assumed that the diastolic blood pressure is smaller, on average, on the left hand of men than on their right hand. To check this assumption, the blood pressure was measured in a sample of 8 healthy men of similar age. Table 3.8 presents the measurement results.
   Use the data in this table to check the null hypothesis that the mean blood pressure on both hands is the same.

*Hint:* Compute the differences "right hand pressure minus left hand pressure" and construct a confidence interval for the mean value of these differences.

**4.** The paper by Bisgaard (2002) presents data on the tensile strength of high voltage electric cables. Each cable is composed of 12 wires. To examine the tensile strength, nine cables were sampled from those produced. For each cable, a small piece of each of the 12 wires was subjected to a tensile strength test. The data are shown in Table 3.9.[1] The box and whisker plot for these data is shown in Fig. 3.4.
   **a.** Compute the cable tensile strength variances and use Hartley's test to check the null hypothesis that all nine sample are drawn from populations with equal variances.

---

[1] Reprinted from *Quality Engineering* (2002), Volume 14(4), p. 680, by courtesy of Marcel Dekker, Inc.

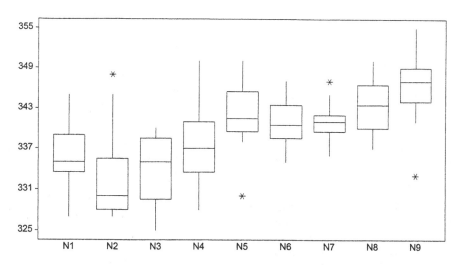

Figure 3.4. Box and whisker plot for cable tensile strength data

*Hint:* $\max s_i^2 = s_2^2 = 48.4, \min s_i^2 = s_7^2 = 10.1$. Hartley's test statistic is 4.8, which is smaller than the 0.05-critical level for $k = 9$ samples and $n - 1 = 11$ which lies between 8.28 and 6.72; see Table 3.7.

**b.** Use the Kruskal–Wallis procedure to test the null hypothesis that the mean strength for all nine cables is the same.

*Solution.* The average ranks for samples 1 through 9 are as shown below:

$$\overline{R}_1 = 36.3, \overline{R}_2 = 26.9, \overline{R}_3 = 28.3, \overline{R}_4 = 44.3, \overline{R}_5 = 67.6,$$
$$\overline{R}_6 = 61.4, \overline{R}_7 = 63.2, \overline{R}_8 = 74.5, \overline{R}_9 = 88.0.$$

The Kruskal–Wallis statistic (3.2.4) is equal to 45.4, which is far above the 0.005 critical value for $\nu = I - 1 = 8$ degrees of freedom. Thus, we definitely reject the null hypothesis.

**c.** Using multiple comparisons, determine groups of cables which are similar with respect to their mean tensile strength.

*Solution.* Statistix produces the following result at $\alpha = 0.05$:

There are 3 groups in which the means are not significantly different from one another.

Group 1: 9,8,7,6,5;
Group 2: 8,5,7,6,4,1;
Group 3: 1,2,3,4,5,6,7.

Bisgaard (2002) notes "that a more detailed examination of the manufacturing process revealed that the cables had been manufactured from raw materials

Table 3.9: Tensile strength of 12 wires for each of nine cables (Bisgaard 2002, p. 680)

| Wire no | 1 | 2 | 3 | 4 | 5 | 6 | 7 | 8 | 9 |
|---|---|---|---|---|---|---|---|---|---|
| 1 | 345 | 329 | 340 | 328 | 347 | 341 | 339 | 339 | 342 |
| 2 | 327 | 327 | 330 | 344 | 3410 | 340 | 340 | 340 | 346 |
| 3 | 335 | 332 | 325 | 342 | 345 | 335 | 342 | 347 | 347 |
| 4 | 338 | 348 | 328 | 350 | 340 | 336 | 341 | 345 | 348 |
| 5 | 330 | 337 | 338 | 335 | 350 | 339 | 336 | 350 | 355 |
| 6 | 334 | 328 | 332 | 332 | 346 | 340 | 342 | 348 | 351 |
| 7 | 335 | 328 | 335 | 328 | 345 | 342 | 347 | 341 | 333 |
| 8 | 340 | 330 | 340 | 340 | 342 | 345 | 345 | 342 | 347 |
| 9 | 337 | 345 | 336 | 335 | 340 | 341 | 341 | 337 | 350 |
| 10 | 342 | 334 | 339 | 337 | 339 | 338 | 340 | 346 | 347 |
| 11 | 333 | 328 | 335 | 337 | 330 | 346 | 336 | 340 | 348 |
| 12 | 335 | 330 | 329 | 340 | 338 | 347 | 342 | 345 | 341 |

taken from two different lots, cable Nos. 1–4 having been made from lot A and cable Nos. 5–9 from lot B".

# Chapter 4

# Sources of Uncertainty: Process and Measurement Variability

*Happy the man, who, studying Nature's laws
through known effects can trace the secret cause.*

Virgil

## 4.1  Introduction: Sources of Uncertainty

Suppose that we have a production process which is monitored by taking samples and measuring the parameters of interest. The results of these measurements are not constant, they are subject to uncertainty. Two main factors are responsible for this uncertainty: changes in the process itself (process variability) and measurement errors (measurement process variability). Suppose, for example, that we are interested in controlling the magnesium content in steel rods. In the normal course of rod production, the magnesium content will vary due to variations in the chemical content of raw materials, "normal" deviations in the parameters of the technological process, temperature variations, etc. So, even an "ideal" laboratory which does absolutely accurate measurements would obtain variable results. In real life there is no such thing as an "ideal" measurement laboratory. Suppose that we prepare several specimens which have practically the same magnesium content. This can be done, for example, by crushing the steel into powder and subsequent mixing. The results of the chemical analysis for magnesium content will be different for different samples. The variability is introduced by the operator, measurement instrument bias and "pure" measurement errors, i.e. the uncertainty built into the chemical measurement procedure.

In this chapter we describe several statistical models which allow us to estimate separately the two main contributions to the overall variability: production process variability and measurement process variability. The first model described in the next section is a one-way ANOVA with random effects.

## 4.2    Process and Measurement Variability: One-way ANOVA with Random Effects

*Example 4.2.1: Titanium content in heat resistant steel*

A factory produces a heat-resistant steel alloy with high titanium (Ti) content. To monitor the process, each day several specimens of the alloy are prepared from the production line. Seven samples are prepared from these specimens and sent for analysis to the local laboratory to establish the Ti content. The experiment is designed and organized in such a way that all seven samples from the daily production must have theoretically the same Ti content.

The results are tabulated in Table 4.1 and a box and whisker plot is given in Fig. 4.1. One can see that there are variations between the samples on a given day, and that there is a considerable day-to-day variability of the average Ti content.

The production manager claims that the variations in the measurement results on each of the five days prove that the measurement process is unstable and erroneous. The laboratory chief, on the other hand, claims that the measurements are quite accurate and that the Ti content varies because the production process is not stable enough. As evidence he points to the day-to-day variations in the average Ti content.

Theoretically, in the absence of measurement errors, all seven samples taken from a single day's batch must have the same percentage of titanium, and if the process itself is stable, the day-to-day variation of titanium content must also be very small.

The purpose of our analysis is the investigation of the day-to-day variability and the variability due to the uncertainty introduced by the measurement process. Our analysis is in fact an application of so-called one-way random factor (or random effects) analysis of variance (ANOVA). It is very important to use accurate notation and to use the appropriate probability model to describe the variation in the results.

Table 4.1: Ti contents of steel rods ($\times 0.01\%$)

| Sample No. | Day 1 | Day 2 | Day 3 | Day 4 | Day 5 |
|---|---|---|---|---|---|
| 1 | 180 | 172 | 172 | 183 | 173 |
| 2 | 178 | 178 | 162 | 188 | 163 |
| 3 | 173 | 170 | 160 | 167 | 168 |
| 4 | 178 | 178 | 163 | 180 | 170 |
| 5 | 173 | 165 | 165 | 173 | 157 |
| 6 | 175 | 170 | 165 | 172 | 174 |
| 7 | 173 | 177 | 153 | 180 | 171 |
| Day aver. | 175.71 | 172.86 | 162.86 | 177.57 | 168.00 |
| | $s_1 = 2.93$ | $s_2 = 4.98$ | $s_3 = 5.76$ | $s_4 = 7.23$ | $s_5 = 6.06$ |

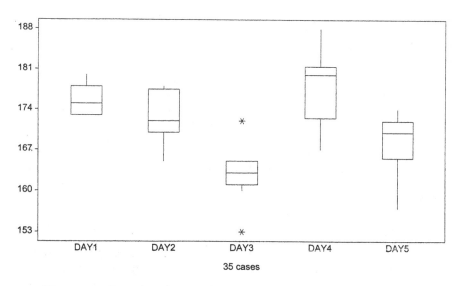

Figure 4.1. The box and whisker plot for the data in Example 4.2.1

## The Model

Let $i$ be the day (batch) number, $i = 1, 2, \ldots, I$, and let $j$ be the measurement number (i.e. the sample number) within a batch, $j = 1, 2, \ldots, J$. $X_{ij}$ denotes the

Table 4.2: The data in matrix form

| Sample No | Batch 1 | Batch 2 | ... | Batch $I$ |
|-----------|---------|---------|-----|-----------|
| 1 | $x_{11}$ | $x_{21}$ | ... | $x_{I1}$ |
| 2 | $x_{12}$ | $x_{22}$ | ... | $x_{I2}$ |
| . | ... | ... | ... | ... |
| . | ... | ... | ... | ... |
| . | ... | ... | ... | ... |
| $J$ | $x_{1J}$ | $x_{2J}$ | ... | $x_{IJ}$ |
|   | $\bar{x}_1.$ | $\bar{x}_2.$ | ... | $\bar{x}_I.$ |

Here $A_i$ is the *random* batch deviation from the overall mean. It changes randomly from batch to batch (from day to day), but remains constant for all samples within one day, i.e. $A_i$ remains constant for all measurements made on the samples prepared from the material produced during one day. $\epsilon_{ij}$ is the random measurement error.

The model (4.1.1) is the so-called single random effect (or single random factor) model. An excellent source on the ANOVA with random effects is Sahai and Ageel (2000). The mathematical model of a single-factor model with random effects is described there on pp. 11 and 24.

Table 4.2 presents the data in matrix form. The measurement results from batch (day) $i$ are positioned in the $i$th column, and $x_{ij}$ is the measurement result from sample $j$ of batch $i$.

Further analysis is based on the following assumptions:

(i) The $A_i$ are assumed to be randomly distributed with mean zero and variance $\sigma_A^2$.

(ii) The $\epsilon_{ij}$ are assumed to be randomly distributed with mean zero and variance $\sigma_e^2$.

iii) For any pair of observations, the corresponding measurement errors are uncorrelated; any $A_i$ is uncorrelated with any other $A_j, i \neq j$, and with any measurement error $\epsilon_{kl}$.

It follows from (4.2.1) that

$$\text{Var}[X_{ij}] = \sigma_A^2 + \sigma_e^2. \tag{4.2.2}$$

Thus the total variance of the measurement results is a *sum of two variances*, one due to the batch-to-batch variance $\sigma_A^2$ and the other due to the measurement error variance $\sigma_e^2$. The main goal of our analysis is to estimate the components of the variance, $\sigma_A^2$ and $\sigma_e^2$.

### Estimation of $\sigma_A$ and $\sigma_e$

First, define the $i$th batch sample mean

$$\overline{x}_{i.} = \frac{x_{i1} + x_{i2} + \ldots + x_{iJ}}{J},$$

(4.2.3)

and the overall sample mean $\overline{x}_{..}$

$$\overline{x}_{..} = \frac{\overline{x}_{1.} + \ldots + \overline{x}_{I.}}{I}.$$

(4.2.4)

Define the sum of squares of the batch sample mean deviation from the overall mean, $ss_A$, according to the following formula:

$$ss_A = J \sum_{i=1}^{I} (\overline{x}_{i.} - \overline{x}_{..})^2.$$

(4.2.5)

This is the *between-batch* variation.

The second principal quantity is the *within-batch* variation $ss_e$:

$$ss_e = \sum_{i=1}^{I} \sum_{j=1}^{J} (x_{ij} - \overline{x}_{i.})^2 = (J-1)(s_1^2 + \ldots + s_I^2),$$

(4.2.6)

where

$$s_i^2 = \sum_{j=1}^{J} (x_{ij} - \overline{x}_{i.})^2 / (J-1).$$

(4.2.7)

Now we are ready to present the formulas for the point estimates of $\sigma_e$ and $\sigma_A$:

$$\hat{\sigma}_e = \sqrt{\frac{ss_e}{I(J-1)}}.$$

(4.2.8)

$$\hat{\sigma}_A = \sqrt{\frac{ss_A/(I-1) - \hat{\sigma}_e^2}{J}}$$

(4.2.9)

If the expression under the square root in (4.2.9) is negative, the estimate of $\sigma_A$ is set to zero.

Let us outline the theory behind these estimates. Let $\overline{X}_{i.}$ be the random variable expressing the average for day $i$, $i = 1, \ldots, I$:

$$\overline{X}_{i.} = \frac{X_{i1} + X_{i2} + \ldots + X_{iJ}}{J}.$$

(4.2.10)

Denote

$$\overline{X}_{..} = (\overline{X}_{1.} + \ldots + \overline{X}_{I.})/I,$$

(4.2.11)

Table 4.3: Analysis of variance for model (4.2.1)

| Source of variation | Degrees of freedom | Sum of squares | Mean square | Expected mean square | F-value |
|---|---|---|---|---|---|
| Between | $I-1$ | $SS_A$ | $MS_A$ | $\sigma_e^2 + J\sigma_A^2$ | $MS_A/MS_e$ |
| Within | $I(J-1)$ | $SS_e$ | $MS_e$ | $\sigma_e^2$ | |
| Total | $IJ-1$ | $SS_T$ | | | |

the mean of all observations. Denote by $MS_A$ and $MS_e$ the so-called mean squares:

$$MS_A = SS_A/(I-1), \ MS_e = SS_e/(I(J-1)). \tag{4.2.12}$$

Then it can be proved that the *expectation* of the mean squares is equal to

$$E[MS_e] = E[\sum_{i=1}^{I}\sum_{j=1}^{J}(X_{ij} - \overline{X}_{i\cdot})^2/I(J-1)] = \sigma_e^2 \tag{4.2.13}$$

and

$$E[MS_A] = E[\sum_{i=1}^{I}\sum_{j=1}^{J}(\overline{X}_{i\cdot} - \overline{X}_{\cdot\cdot})^2/(I-1)] = \sigma_e^2 + J\sigma_A^2. \tag{4.2.14}$$

Table 4.3 summarizes this information

Point estimates $\hat{\sigma}_e$ and $\hat{\sigma}_A$ in (4.2.8), (4.2.9) are obtained by replacing $X_{ij}, \overline{X}_{i\cdot}$ and $\overline{X}_{\cdot\cdot}$ by the observed values $x_{ij}, \overline{x}_{i\cdot}$ and $\overline{x}_{\cdot\cdot}$, respectively. Section 2.5 of Sahai and Ageel (2000) contains more details on the point estimation of $\sigma_A$ and $\sigma_e$.

It is important to mention that (4.2.13) and (4.2.14) are true without assuming any specific form of a distribution regarding the random variables $A_i$ and $\epsilon_{ij}$.

## Example 4.2.1 concluded.  Additional statistical checks

Let us complete the computations for the titanium example. Here $I = 5$ and $J = 7$. From Table 4.1 it follows that

$$\overline{x}_{\cdot\cdot} = (175.71 + 172.86 + 162.86 + 177.57 + 168.00)/5 = 171.40, \tag{4.2.15}$$

$$ss_A = 7\left((175.71 - 171.40)^2 + \ldots + (168.00 - 171.40)^2\right) = 1002.9, \tag{4.2.16}$$

$$ss_e = 6(2.93^2 + 4.98^2 + 5.76^2 + 7.23^2 + 6.06^2) = 933.4. \tag{4.2.17}$$

Table 4.4: Optimal number of measurements $J^*$ per batch for $N = 60$

| $\theta = 0.025$ | $\theta = 0.05$ | $\theta = 0.1$ | $\theta = 0.2$ | $\theta = 0.4$ | $\theta = 0.8$ | $\theta = 1.0$ |
|---|---|---|---|---|---|---|
| 30 | 12 | 10 | 6 | 4 | 3 | 3 |

Now by (4.2.8) and (4.2.9),

$$\hat{\sigma}_e = \sqrt{933.4/30} = 5.58 \approx 5.6; \qquad (4.2.18)$$

$$\hat{\sigma}_A = \sqrt{(1002.9/4 - 5.58^2)/7} = 5.6. \qquad (4.2.19)$$

We see, therefore, that the measurement errors and day-to-day variability of the production process make equal contributions to the total variability of measurement results.

If the computation results show that $\sigma_A$ is very small, then it is a good idea to test the null hypothesis that $\sigma_A = 0$. The corresponding testing procedure is carried out under the normality assumption regarding all random variables involved, $A_i$ and $\epsilon_{ij}$. This procedure is based on the so-called $F$-ratio defined in Table 4.3, and it works as follows.

Compute the ratio

$$F = \frac{ss_A/(I - 1)}{ss_e/(I(J - 1))} \qquad (4.2.20)$$

If $F$ *exceeds* the $\alpha$-critical value of the $F$-statistic with $I - 1, I(J - 1)$ degrees of freedom, $\mathcal{F}_{I-1,I(J-1)}(\alpha)$, then the null hypothesis $\sigma_A = 0$ is rejected at the significance level $\alpha$ in favor of the alternative $\sigma_A > 0$.

For example 4.2.1, $F = 8.06$. From Appendix C we see that for $\alpha = 0.01$, $\mathcal{F}_{4,30}(0.01) = 4.02$. Since this number is *smaller* than the computed $F$-value, we reject the null hypothesis at level $\alpha = 0.01$.

## Optimal Number of Measurements per Batch

The following issue is of theoretical and practical interest. Suppose that we decide to make in total $N$ measurements, where $N = I \times J$, $I$ being the number of batches and $J$ being the number of measurements per batch. Suppose that our purpose is the estimation with maximal precision (i.e. with minimal variance) of the ratio $\theta = \sigma_A^2/\sigma_e^2$. The variance in estimating $\theta$ depends on the choice of $I$ and $J$ (subject to the constraint on their product $I \times J = N$). Table 4.4 shows the optimum number of measurements per batch $J^*$ depending on the assumed value of $\theta$, for $N = 60$.

Returning to Example 4.2.1, $\theta \approx 1$. The best allocation of $N = 60$ experiments would be to take $J^* = 3$ measurements per batch and to take 20 batches.

The general rule is the following: If $\sigma_e \gg \sigma_A$, i.e. $\theta$ is small, it is preferable to take many observations per batch and a few batches. If $\sigma_A \approx \sigma_e$, it is preferable to take many batches, with few observations per batch. Suppose that for our example we choose $I \times J = 36$. The optimal allocation design would be to take $I = 3$ batches and 12 measurements per batch.

### Concluding remarks

In the model considered in this section, apart from the measurement errors, the only additional source of result variability is the process batch-to-batch variation. This was reflected in choice of model, the one-factor ANOVA with random effects. In production and measurement practice, we meet more complicated situations. Assume, for example, that the batches are highly nonhomogeneous, and each batch is divided into several *samples* within which the product is expected to have relatively small variations. Formalization of this situation leads to a two-factor hierarchical (or nested) design. This will be considered in next section.

It happens quite frequently that there are several, most often two, principal factors influencing the measurement process variability. A typical situation with two random factors is described in Sect. 4.4. There we will consider a model with two random sources influencing measurement result variability, one of which is the part-to-part variation, and the other the variability brought into the measurement process by using different operators to carry out the measurements.

## 4.3   Hierarchical Measurement Design

Our assumption in Example 4.2.1 was that a single batch is completely homogeneous with regard to the titanium content. This, in fact, is valid only in rather special circumstances. For example, each day's production consists of several separately produced portions (e.g. each portion is produced on a different machine), which then go through a mechanical mill and are thoroughly mixed together. So, each batch given to the lab is *homogenized*. This is a clever technique which helps to avoid machine-to-machine variability. However, the production process does not always allow such homogenization and/or it is in our interest to discover and to evaluate the variability introduced by different machines.

Consider, for example, the following situation. Each day a long metal rod is produced; along its length there might be some variations in titanium content. We might be interested in estimating these variations and in separating them from the daily batch-to-batch variations.

The following design of experiment allows us to investigate the contribution to the total variability of the following three sources: the day-to-day variation; the sample-to-sample variation within one day; and the variability caused by measurement errors.

We will refer to each day's output as a batch. In our example the batch will be a long metal rod. We analyze $I$ batches. For the analysis, $J$ samples are randomly taken from each batch. In our case it might be a collection of $J$ small specimens cut randomly along the whole length of the rod. Each sample is measured $K$ times in the lab to determine its titanium content. The variation in these $K$ measurements from one sample is introduced solely by the variations in the measurement process. If the measurement process were "ideal", all $K$ measurements would be identical.

Schematically, this measurement design can be presented in the form of a "tree", as shown on Fig. 4.2. In the literature it is called often a "nested" or "hierarchical" design.

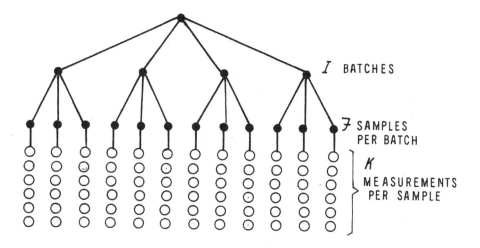

Figure 4.2. A hierarchical design of measurements

## 4.3.1 The Mathematical Model of Hierarchical Design

We number the batches $i = 1, \ldots, I$, the samples within each batch $j = 1, \ldots, J$, and the measurements (tests) within each sample $k = 1, 2, \ldots, K$. $X_{ijk}$ is the measurement result of the $k$th measurement in the $j$th sample of the $i$th batch.

We assume that

$$X_{ijk} = \mu + A_i + B_{j(i)} + \epsilon_{ijk}, \tag{4.3.1}$$

where $\mu$ is the overall mean value, $A_i$ represents the random contribution to $\mu$ due to $i$th batch, $B_{j(i)}$ is the random contribution to $\mu$ due to sample $j$ nested within batch $i$, and $\epsilon_{ijk}$ is the error term that reflects the measurement error for the $k$th measurement of the $j$th sample in the $i$th batch.

We make the following assumptions:

(i) The $A_i$ are zero-mean random variables with variance $\sigma_A^2$.

(ii) The $B_{j(i)}$ are zero-mean random variables with variance $\sigma_B^2$.

(iii) The $\epsilon_{ijk}$ are zero-mean random variables with variance $\sigma_e^2$.

(iv) The $A_i$ and $B_{j(i)}$ are mutually uncorrelated; the $\epsilon_{ijk}$ are uncorrelated between themselves and uncorrelated with $A_i$ and $B_{j(i)}$.

Our purpose is to estimate $\sigma_e, \sigma_A$ and $\sigma_B$. We need some notation. The observed value of the random variable $X_{ijk}$ will be denoted as $x_{ijk}$. Let $\bar{x}_{ij\cdot}$ be the sample average of all $K$ measurements in the $j$th sample of batch $i$:

$$\bar{x}_{ij\cdot} = (x_{ij1} + \ldots + x_{ijK})/K. \tag{4.3.2}$$

Let $\bar{x}_{i\cdot\cdot}$ be the sample average of all measurements in the $i$th batch:

$$\bar{x}_{i\cdot\cdot} = (\bar{x}_{i1\cdot} + \ldots + \bar{x}_{iJ\cdot})/J. \tag{4.3.3}$$

Let $\bar{x}_{\cdots}$ be the overall average:

$$\bar{x}_{\cdots} = (\bar{x}_{1\cdot\cdot} + \ldots + \bar{x}_{I\cdot\cdot})/I = \sum_{i=1}^{I}\sum_{j=1}^{J}\sum_{k=1}^{K} x_{ijk}. \tag{4.3.4}$$

We will need the following three sums of squares:

$$ss_e = \sum_{i=1}^{I}\sum_{j=1}^{J}\sum_{k=1}^{K}(x_{ijk} - \bar{x}_{ij\cdot})^2, \tag{4.3.5}$$

which is called the sum of squares for the pure measurement error;

$$ss_{B(A)} = K\sum_{i=1}^{I}\sum_{j=1}^{J}(\bar{x}_{ij\cdot} - \bar{x}_{i\cdot\cdot})^2, \tag{4.3.6}$$

the so-called between-sample sum of squares, and

$$ss_A = K \cdot J \sum_{i=1}^{I}(\bar{x}_{i\cdot\cdot} - \bar{x}_{\cdots})^2, \tag{4.3.7}$$

the between-batch sum of squares. Here are the formulas for $\hat{\sigma}_e, \hat{\sigma}_A$ and $\hat{\sigma}_B$:

$$\hat{\sigma}_e = \sqrt{\frac{ss_e}{IJ(K-1)}}; \tag{4.3.8}$$

$$\hat{\sigma}_B = \sqrt{\left(\frac{ss_{B(A)}}{I(J-1)} - \hat{\sigma}_e^2\right)/K}; \tag{4.3.9}$$

$$\hat{\sigma}_A = \sqrt{\left(\frac{ss_A}{I-1} - \hat{\sigma}_e^2 - K \cdot \hat{\sigma}_B^2\right)/J \cdot K}. \tag{4.3.10}$$

Table 4.5: Analysis of variance for model (4.3.1)

| Source of variation | Degree of freedom | Sum of squares | Mean square | Expected mean square |
|---|---|---|---|---|
| Due to $A$ | $I-1$ | $SS_A$ | $MS_A$ | $\sigma_e^2 + K\sigma_B^2 + JK\sigma_A^2$ |
| $B$ within $A$ | $I(J-1)$ | $SS_{B(A)}$ | $MS_{B(A)}$ | $\sigma_e^2 + K\sigma_B^2$ |
| Error | $IJ(K-1)$ | $SS_e$ | $MS_e$ | $\sigma_e^2$ |
| Total | $IJK-1$ | $SS_T$ | | |

Note that if the expression under the square root in (4.3.9) or (4.3.10) is negative, then the corresponding estimate is set to zero. Table 4.5 summarizes the information used for point estimation of $\sigma_e, \sigma_A$ and $\sigma_B$.

It is worth noting that the formulas for point estimates (4.3.8)–(4.3.10) are derived without using any assumptions regarding the form of the distribution of random variables $A_i, B_{i(j)}, \epsilon_{ijk}$. The normality of their distribution must be assumed at the stage of testing hypotheses regarding the parameters involved.

*Example 4.3.1: Hierarchical measurement scheme* [1]
The data in Table 4.6 are borrowed from Box (1998, p. 174). We have $I = 5$ batches, $J = 2$ samples per batch and $K = 2$ measurements per sample. To compute the first sum of squares, $ss_e$, let us recall a useful shortcut: if there are two observations in a sample, say $x_1$ and $x_2$, then the sum of squares of the deviations from their average is $(x_1 - x_2)^2/2$. Thus

$$ss_e = \frac{(74.1 - 74.3)^2}{2} + \ldots + \frac{(78.2 - 78.0)^2}{2} = 0.98$$

Then by (4.3.8), $\hat{\sigma}_e = \sqrt{0.98/10} = 0.313 \approx 0.31$.

Since there are two samples per one batch, we can apply the same shortcut to compute $ss_{B(A)}$:

$$ss_{B(A)} = 2 \cdot \left( \frac{(74.2 - 68.0)^2}{2} + \ldots + \frac{(81.9 - 78.1)^2}{2} \right) = 234.0$$

and thus by (4.3.9), $\hat{\sigma}_B = 4.83 \approx 4.8$.
The batch averages are

$\bar{x}_{1..} = (74.2 + 68.0)/2 = 71.1,$
$\bar{x}_{2..} = (75.1 + 71.5)/2 = 73.3,$
$\bar{x}_{3..} = (59.0 + 63.4)/2 = 61.2,$
$\bar{x}_{4..} = (82.0 + 69.8)/2 = 75.9,$

[1]Reprinted from *Quality Engineering* (1998-99), Volume 11(1), p. 174, by courtesy of Marcel Dekker, Inc.

Table 4.6: Titanium content ×0.01%

| Batch | 1 | 1 | 2 | 2 | 3 | 3 | 4 | 4 | 5 | 5 |
|---|---|---|---|---|---|---|---|---|---|---|
| Sample | 1 | 2 | 1 | 2 | 1 | 2 | 1 | 2 | 1 | 2 |
| | 74.1 | 68.2 | 75.4 | 71.5 | 59.4 | 63.2 | 81.7 | 69.9 | 81.7 | 78.2 |
| | 74.3 | 67.8 | 74.8 | 71.5 | 58.6 | 63.6 | 82.3 | 69.7 | 82.1 | 78.0 |
| Average | 74.2 | 68.0 | 75.1 | 71.5 | 59.0 | 63.4 | 82.0 | 69.8 | 81.9 | 78.1 |

$\bar{x}_{5..} = (81.9 + 78.1)/2 = 80.0,$

and the overall average (mean) is

$(71.1 + 73.3 + 61.2 + 75.9 + 80.0)/5 = 72.3.$

Then, by (4.3.7),

$$ss_A = 2 \cdot 2\Big((71.1 - 72.3)^2 + \ldots + (80.0 - 72.3)^2\Big) = 791.6.$$

Now by (4.3.10),

$$\hat{\sigma}_A = \sqrt{(791.6/4 - 0.098 - 2 \times 4.83^2)/(2 \times 2)} = 6.14 \approx 6.1.$$

The total variance of $X_{ijk}$ is $\text{Var}[X_{ijk}] = \sigma_A^2 + \sigma_B^2 + \sigma_e^2$. Its estimate is $6.14^2 + 4.83^2 + 0.31^2 = 61.1$. The greatest contribution to this quantity is due to the batch-to batch variation, about 62%. The second largest is the sample-to-sample variability within one batch, about 38%. The variability due to measurement errors gives a negligible contribution to the total variability, about 0.2%.

## 4.3.2   Testing the Hypotheses $\sigma_A = 0, \sigma_{B(A)} = 0$

Hypothesis testing in the random effects model rests on the assumption that all random variables involved are normally distributed.

To test the null hypothesis $\sigma_A^2 = 0$ against alternative $\sigma_A^2 > 0$ it is necessary to compute the statistic

$$F_A = \frac{ss_A/(I - 1)}{ss_{B(A)}/I(J - 1)}. \tag{4.3.11}$$

The null hypothesis is *rejected* at significance level $\alpha$ if the computed value of $F_A$ exceeds $\mathcal{F}_{I-1,I(J-1)}(\alpha)$, which is the $\alpha$-critical value of the $\mathcal{F}$ distribution with $\nu_1 = I - 1$ and $\nu_2 = I(J - 1)$ degrees of freedom. These critical values are presented in Appendix C.

Similarly, to test the null hypothesis $\sigma_B^2 = 0$ against alternative $\sigma_B^2 > 0$, it is necessary to compute the statistic

$$F_B = \frac{ss_{B(A)}/I(J - 1)}{ss_e/IJ(K - 1)}. \tag{4.3.12}$$

The null hypothesis is *rejected* at significance level $\alpha$ if the computed value of $F_B$ exceeds $\mathcal{F}_{I(J-1),IJ(K-1)}(\alpha)$, which is the $\alpha$-critical value of the $\mathcal{F}$ distribution with $\nu_1 = I(J-1)$ and $\nu_2 = IJ(K-1)$ degrees of freedom.

As an exercise, check that for Example 4.3.1 the null hypotheses for $\sigma_A^2$ and $\sigma_B^2$ are rejected at significance level 0.1 and 0.05, respectively.

# 4.4 Repeatability and Reproducibility Studies

## 4.4.1 Introduction and Definitions

The 1994 edition of US National Institute of Standards and Technology (NIST) Technical Note 1297, *Guidelines for Evaluating and Expressing the Uncertainty of NIST Measurement Results*, defines the *repeatability* of measurement results as "closeness of the agreement between the results of successive measurements of the same measurand carried out under the same conditions of measurement". It goes on to say: "These conditions, called *repeatability conditions*, include the same measurement procedure, the same observer, the same measuring instrument, under the same conditions, the same location, repetition over a short period of time" (clause D.1.1.2). Repeatability may be expressed quantitatively in terms of the dispersion (variance) characteristics of the measurement results carried out under repeatability conditions.

The *reproducibility* of measurement results is defined as "closeness of the agreement between the results of measurements of the same measurand carried out under changed conditions of measurement". The changed conditions may include "principle of measurement, method of measurement, different observer, measuring instrument, location, conditions of use, time. Reproducibility may be expressed quantitatively in terms of the dispersion (variance) characteristics of the results" (clause D.1.1.3).

In all our previous examples, we spoke about "measurements" without specifying the exact measurement conditions, tools and methods. The reader probably understood that all measurements were carried out on a single measurement device, by a single operator, within a short period of time and under permanent, unchanged conditions. In practice, however, all these limitations are rarely satisfied. Measurements are performed on several measurement devices, by several different operators, over a protracted period of time and under variable conditions. All that increases of course the variability of the results. Repeatability and reproducibility (R&R) studies are meant to investigate the contribution of various factors to the variability of measurement results. While in Sects. 4.2 and 4.3 our attention was focused on separating the process variability from the variability introduced by the measurements, in this section we focus on the *measurement process* and study various sources of variability in the measurement process itself. Often in literature the subject of this chapter is called "Gauge capability studies." The word "gauge" means a measuring tool or a means of making an estimate or judgment.

Table 4.7: Measured weights (in grams) of $I = 2$ pieces of paper taken by $J = 5$ operators

| Item | Operator 1 | Operator 2 | Operator 3 | Operator 4 | Operator 5 |
|------|------------|------------|------------|------------|------------|
| 1 | 3.481 | 3.448 | 3.485 | 3.475 | 3.472 |
| 1 | 3.477 | 3.472 | 3.464 | 3.472 | 3.470 |
| 1 | 3.470 | 3.470 | 3.477 | 3.473 | 3.474 |
| 2 | 3.258 | 3.254 | 3.256 | 3.249 | 3.241 |
| 2 | 3.254 | 3.247 | 3.257 | 3.238 | 3.250 |
| 2 | 3.258 | 3.239 | 3.245 | 3.240 | 3.254 |

## 4.4.2  Example of an R&R Study

*Example 4.4.1: Weight of paper measured by five operators*
This example, borrowed from Vardeman and VanValkenburg (1999), presents a typical framework for R&R studies. Five operators, chosen randomly from the pool of operators available, carried out weight measurements of two items (two pieces of paper). Each operator took three measurements of each item). The measured weights in grams are presented in Table 4.7.[2]

There are four sources of the result variability: item, operators, item–operator interactions and measurements. The variation within one cell in the Table 4.7, where the item, and the operator are the same, reflects the repeatability. The variation due to the operator and the item–operator interaction reflects the reproducibility.

Figure 4.3 illustrates the data structure in general form. It is convenient to represent the measurement data in a form of a matrix. This has $I$ rows, $J$ columns and $I \times J$ cells. Here $I$ is the number of items, $J$ is the number of operators. The $(i, j)$th cell contains $K$ measurements carried out by the same operator on the same item. In Example 4.4.1, $I = 2$, $J = 5$, $K = 3$.

**The Model**

To give the notions of repeatability and reproducibility more accurate definitions, we need a mathematical model describing the formal structure of the measurement results.

The $k$th measurement result in the $(i, j)$th cell is considered as a random variable which will be denoted as $X_{ijk}$. We assume that it has the following form:

$$X_{ijk} = \mu + A_i + B_j + (AB)_{ij} + \epsilon_{ijk}. \tag{4.4.1}$$

---

[2] Reprinted with permission from *Technometrics*. Copyright 1999 by the American Statistical Association. All rights reserved.

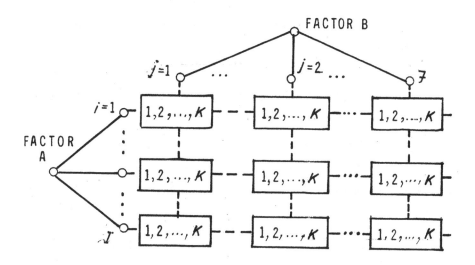

Fig. 4.3. Graphical representation of a two–factor balanced design model

Here $\mu$ is the overall mean; $A_i$ is the random contribution to $\mu$ due to item $i$; $B_j$ is the random contribution to $\mu$ due to operator $j$; $(AB)_{ij}$ is the random term describing the joint contribution to $\mu$ of item $i$ and operator $j$, which is called the random $i - j$ interaction term. $\epsilon_{ijk}$ is the "pure" measurement error in the $k$th measurement for a fixed combination of $i$ and $j$.

Formally (4.4.1) is called a two-way crossed design with random factors (or random effects); see Sahai and Ageel (2000, Chapt. 4).

The observed values of $X_{ijk}$ will be denoted as $x_{ijk}$. For example, the observed $K$ measurement results in the $(i, j)$th cell will be denoted as $x_{ij1}, \ldots, x_{ijK}$.

We make the following assumptions regarding the random variables $A_i, B_j$, $(AB)_{ij}$ and $\epsilon_{ijk}$:

   (i) The $A_i$ are zero-mean variables with variance $\sigma_A^2$.
   (ii) The $B_j$ are zero-mean random variables with variance $\sigma_B^2$.
   (iii) The $(AB)_{ij}$ are zero-mean random variables with variance $\sigma_{AB}^2$.
   (iv) The $\epsilon_{ijk}$ are zero-mean random variables with variance $\sigma_e^2$.
   (v) All the above-mentioned random variables are mutually uncorrelated.
Assumptions (i)–(v) allow us to obtain point estimates of the variances involved and of some functions of them which are of interest in repeatability and reproducibility studies.

To obtain confidence intervals and to test hypotheses regarding the variances, the following assumption will be made:

   (vi) All random variables involved, $A_i, B_j, (AB)_{ij}$ and $\epsilon_{ijk}$, are normally distributed.

Repeatability is measured by $\sigma_e$. We use the notation $\sigma_{repeat} = \sigma_e$. Reproducibility is measured by the square root of the variance components $\sigma_B^2 + \sigma_{AB}^2$. The quantity

$$\sigma_{repro} = \sqrt{\sigma_B^2 + \sigma_{AB}^2} \tag{4.4.2}$$

is the standard deviation of measurements made by many operators measuring the same item in the absence of repeatability variation. In R&R studies, an important parameter is

$$\sigma_{R\&R} = \sqrt{\sigma_{repeat}^2 + \sigma_{repro}^2} = \sqrt{\sigma_e^2 + \sigma_B^2 + \sigma_{AB}^2}, \tag{4.4.3}$$

which reflects the standard deviations of all measurements made on the same item. Finally, another parameter of interest, which is analogous to $\sigma_{repro}$, is

$$\sigma_{items} = \sqrt{\sigma_A^2 + \sigma_{AB}^2}. \tag{4.4.4}$$

## Point Estimation of $\sigma_{repeat}$, $\sigma_{repro}$, $\sigma_{R\&R}$ and $\sigma_{items}$

As already mentioned, there is no need for assumption (vi) in obtaining the point estimates.

The observed measurement results in the $(i, j)$th cell will be denoted as $x_{ijk}, k = 1, \ldots, K$. $X_{ijk}$ will denote the corresponding random measurement result. The sample mean of one cell is denoted as $\bar{x}_{ij.}$:

$$\bar{x}_{ij.} = \sum_{k=1}^{K} x_{ijk}/K. \tag{4.4.5}$$

The sample mean of all cell averages in row $i$ is denoted as $\bar{x}_{i..}$:

$$\bar{x}_{i..} = (\bar{x}_{i1.} + \ldots + \bar{x}_{iJ.})/J. \tag{4.4.6}$$

By analogy, the average of all cell means in the $j$th column is denoted as $\bar{x}_{.j.}$:

$$\bar{x}_{.j.} = (\bar{x}_{1j.} + \ldots + \bar{x}_{Ij.})/I. \tag{4.4.7}$$

The observed value of the mean of all cell means is denoted as $\bar{x}_{...}$:

$$\bar{x}_{...} = \sum_{j=1}^{J} \bar{x}_{.j.}/J = \sum_{i=1}^{I} \bar{x}_{i..}/I. \tag{4.4.8}$$

First, define the following four sums of squares based on the *observed* values of the random variables $X_{ijk}$:

$$ss_A = J \cdot K \cdot \sum_{i=1}^{I} (\bar{x}_{i..} - \bar{x}_{...})^2, \tag{4.4.9}$$

Table 4.8: Analysis of variance for model (4.4.1)

| Source of variation | Degrees of Freedom | Sum of squares | Mean square | Expected mean square |
|---|---|---|---|---|
| Due to $A$ | $I-1$ | $SS_A$ | $MS_A$ | $\sigma_e^2 + K\sigma_{AB}^2 + JK\sigma_A^2$ |
| Due to $B$ | $J-1$ | $SS_B$ | $MS_B$ | $\sigma_e^2 + K\sigma_{AB}^2 + IK\sigma_B^2$ |
| Interaction $A \times B$ | $(I-1)(J-1)$ | $SS_{AB}$ | $MS_{AB}$ | $\sigma_e^2 + K\sigma_{AB}^2$ |
| Error | $IJ(K-1)$ | $SS_e$ | $MS_e$ | $\sigma_e^2$ |
| Total | $IJK-1$ | $SS_T$ | | |

$$ss_B = I \cdot K \cdot \sum_{j=1}^{J} (\overline{x}_{\cdot j \cdot} - \overline{x}_{\dots})^2, \tag{4.4.10}$$

$$ss_{AB} = K \sum_{i=1}^{I} \sum_{j=1}^{J} (\overline{x}_{ij \cdot} - \overline{x}_{i \cdot \cdot} - \overline{x}_{\cdot j \cdot} + \overline{x}_{\dots})^2, \tag{4.4.11}$$

$$ss_e = \sum_{i=1}^{I} \sum_{j=1}^{J} \sum_{k=1}^{K} (x_{ijk} - \overline{x}_{ij \cdot})^2. \tag{4.4.12}$$

Let us now define four *random* sum of squares $SS_A, SS_B, SS_{AB}$ and $SS_e$ by replacing in (4.4.9)–(4.4.12) the lower-case italic letters $x_{ijk}, \overline{x}_{i \cdot \cdot}, \overline{x}_{\cdot j \cdot}$ and $\overline{x}_{\dots}$ by the corresponding capital italic letters which denote random variables. In this replacement operation, $\overline{X}_{ij \cdot}$ is the random mean of all observations in cell $(i,j)$; $\overline{X}_{i \cdot \cdot}$ is the random mean of all observations in row $i$, $\overline{X}_{\cdot j \cdot}$ is the random mean of all observations in column $j$. $\overline{X}_{\dots}$ denotes the random mean of all $I \times J \times K$ observations. The properties of these random sum of squares are summarized in Table 4.8.

Replacing the expected values of $SS_e, SS_A, SS_B$ and $SS_{AB}$ by their observed values $ss_e, ss_A, ss_B, ss_{AB}$, and the variances $\sigma_e^2, \sigma_A^2, \sigma_B^2, \sigma_{AB}^2$ by their estimates, we arrive at the following four equations for the point estimates of the variances in the model:

$$\frac{ss_e}{IJ(K-1)} = \hat{\sigma}_e^2, \tag{4.4.13}$$

$$\frac{ss_A}{I-1} = \hat{\sigma}_e^2 + K\hat{\sigma}_{AB}^2 + JK\hat{\sigma}_A^2, \tag{4.4.14}$$

$$\frac{ss_B}{J-1} = \hat{\sigma}_e^2 + K\hat{\sigma}_{AB}^2 + IK\hat{\sigma}_B^2, \tag{4.4.15}$$

$$\frac{ss_{AB}}{(I-1)(J-1)} = \hat{\sigma}_e^2 + K\hat{\sigma}_{AB}^2. \tag{4.4.16}$$

Combining the variance estimates, we arrive after little algebra at the following estimates of the parameters which are of interest in R&R studies:

$$\hat{\sigma}_{repro} = \sqrt{\max\left(0, \frac{ss_B}{KI(J-1)} + \frac{ss_{AB}}{IK(J-1)} - \frac{ss_e}{IJ(K-1)K}\right)}. \quad (4.4.17)$$

$$\hat{\sigma}_{R\&R} = \sqrt{\frac{ss_B}{IK(J-1)} + \frac{ss_{AB}}{IK(J-1)} + \frac{ss_e}{KIJ}}. \quad (4.4.18)$$

$$\hat{\sigma}_{items} = \sqrt{\max\left(0, \frac{ss_A}{K(I-1)J} + \frac{ss_{AB}}{JK(I-1)} - \frac{ss_e}{KIJ(K-1)}\right)}. \quad (4.4.19)$$

*Example 4.4.1 continued*
Let us return to the data in Table 4.7 In our case $I = 2, J = 5$ and $K=3$. We omit the routine computations, which we carried out with the two-way ANOVA procedure in Statistix, and present the results:

$$ss_A = 0.37185; \quad\quad\quad\quad\quad\quad\quad\quad\quad\quad\quad\quad (4.4.20)$$

$$ss_B = 0.000502;$$

$$ss_{AB} = 0.000176;$$

$$ss_e = 0.00102.$$

Using formulas (4.4.13) and (4.4.17)–(4.4.19), we find that

$$\hat{\sigma}_{repeat} = 0.0071; \quad\quad\quad\quad\quad\quad\quad\quad\quad\quad\quad (4.4.21)$$

$$\hat{\sigma}_{repro} = 0.0034;$$

$$\hat{\sigma}_{R\&R} = 0.0079;$$

$$\hat{\sigma}_{items} = 0.16.$$

## 4.4.3   Confidence Intervals for R&R Parameters

This subsection contains rather theoretical material, and the reader can skip the theory at first reading and go directly to the formulas presenting the final results.

We will use the following notation: $V \sim \chi^2(\nu)$ means that the random variable $V$ has a chi-square distribution with $\nu$ degrees of freedom. Denote by $q_\beta(\nu)$ the $\beta$-quantile of the chi-square distribution with $\nu$ degrees of freedom. Thus, if $V \sim \chi^2(\nu)$ then

$$P\big(V \le q_{(1-\beta)}(\nu)\big) = 1 - \beta, \quad\quad\quad (4.4.22)$$
$$P\big(V \le q_\beta(\nu)\big) = \beta, \quad\quad\quad (4.4.23)$$
$$P\big(q_\beta(\nu) < V \le q_{(1-\beta)}(\nu)\big) = 1 - 2\beta. \quad\quad\quad (4.4.24)$$

The quantiles for the chi-square distribution are presented in Appendix B for $\beta = 0.01, 0.025, 0.95, 0.975, 0.99$.

All our derivations are based on the following claim proved in statistics courses.

**Claim 4.4.1**

If all random variables in model (4.4.1) have normal distribution, then

$$\frac{SS_A}{\sigma_e^2 + K\sigma_{AB}^2 + JK\sigma_A^2} \sim \chi^2(I-1), \tag{4.4.25}$$

$$\frac{SS_B}{\sigma_e^2 + K\sigma_{AB}^2 + IK\sigma_B^2} \sim \chi^2(J-1), \tag{4.4.26}$$

$$\frac{SS_{AB}}{\sigma_e^2 + K\sigma_{AB}^2} \sim \chi^2\big((I-1) \times (J-1)\big), \tag{4.4.27}$$

$$\frac{SS_e}{\sigma_e^2} \sim \chi^2\big(IJ(K-1)\big). \tag{4.4.28}$$

All $\chi^2(\cdot)$ random variables in (4.4.25)–(4.4.28) are independent.

An immediate application of this claim is the $1 - 2\beta$ confidence interval on $\sigma_e$. It follows from (4.4.26) that

$$P\left(q_\beta(IJ(K-1)) \leq \frac{SS_e}{\sigma_e^2} \leq q_{(1-\beta)}(IJ(K-1))\right) = 1 - 2\beta, \tag{4.4.29}$$

$$P\left(\sqrt{\frac{SS_e}{q_{(1-\beta)}(IJ(K-1))}} \leq \sigma_e \leq \sqrt{\frac{SS_e}{q_\beta(IJ(K-1))}}\right) = 1 - 2\beta. \tag{4.4.30}$$

For the data in Example 4.4.1, $ss_e = 0.00102$, $IJ(K-1) = 20$, $q_{0.025}(20) = 9.591$, $q_{0.975}(20) = 34.17$. By (4.4.29), the 95% confidence interval on $\sigma_e$ is

[0.0055, 0.010].

The confidence interval for $\sigma_{repro}$ will be obtained using the so-called M-method developed in reliability theory (see Gertsbakh 1989, Chap. 5). This method works as follows.

Suppose that the random set $S_{1-2\beta}$ covers in the three-dimensional parametric space the point $\sigma_\star^2 = (\sigma_e^2, \sigma_{AB}^2, \sigma_B^2)$ with probability $1 - 2\beta$. Let $m(S_{1-2\beta})$ and $M(S_{1-2\beta})$ be the minimum and the maximum of the function $\psi(\sigma_\star^2) = 0 \cdot \sigma_e^2 + \sigma_{AB}^2 + \sigma_B^2$ on this set. Then the following implication holds:

$$(\sigma_e^2, \sigma_{AB}^2, \sigma_B^2) \in S_{1-2\beta} \Rightarrow m(S_{1-2\beta}) \leq \psi(\sigma_\star^2) \leq M(S_{1-2\beta}). \tag{4.4.31}$$

Therefore,

$$P\Big(m(S_{1-2\beta}) \leq \psi(\sigma_\star^2) \leq M(S_{1-2\beta})\Big) \geq 1 - 2\beta. \tag{4.4.32}$$

The choice of the confidence set $S_{1-2\beta}$ is most important. We choose it according to (4.4.25) of Claim 4.4.1 for $I = 2, J = 5, K = 3$, $\beta = 0.05$. Note that

$$P\left(SS_B/q_{0.95}(4) \leq \sigma_e^2 + 3\sigma_{AB}^2 + 6\sigma_B^2 \leq SS_B/q_{0.05}(4)\right) = 0.90. \qquad (4.4.33)$$

Thus, the desired 90% confidence set is defined by the inequalities

$$SS_B/q_{0.95}(4) < \sigma_e^2 + 3\sigma_{AB}^2 + 6\sigma_B^2 < SS_B/q_{0.05}(4). \qquad (4.4.34)$$

It is easy to demonstrate that the function $\sigma_{AB}^2 + \sigma_B^2$ has the following maximal and minimal values on this set:

maximal value $M = SS_B/(q_{0.05}(4) \cdot 3)$ and
minimal value $m = SS_B/(q_{0.95}(4) \cdot 6)$.

Therefore, $(m, M)$ is the interval which contains $\sigma_{AB}^2 + \sigma_B^2 = \sigma_{repro}^2$ with probability at least 0.90.

Now substitute $ss_B = 0.000502$, $q_{0.95}(4) = 9.488$, $q_{0.05}(4) = 0.711$ and obtain that the confidence interval for $\sigma_{repro}$ is
$(\sqrt{m}, \sqrt{M}) = [0.003, 0.015]$.

Vardeman and VanValkenburg (1999) suggest constructing the confidence intervals for $\sigma_{R\&R}$ $\sigma_{repro}$ and $\sigma_{items}$ using the approximate values of the standard errors of the corresponding parameter estimates. This method is based on the error propagation formula, and we will explain in the next chapter how this method works.

## 4.5  Measurement Analysis for Destructive Testing

### Pull Strength Repeatability: Experiment Description

We have seen so far that the estimation of measurement repeatability is based on repeated measurements carried out under similar conditions (the same item, the same operator, the same instrument, etc.). There are situations in which it is not possible to repeat the measurements. A typical case is an experiment which destroys the measured specimen. For example, measuring the strength or fatigue limit of a mechanical construction destroys the construction. In order to evaluate the repeatability it is necessary to design the experiment in a special way. The following interesting and instructive example is described by Mitchell et al. (1997).

In a hardware production process, a gold wire is used to connect the integrated circuit (IC) to the lead frame. Ultrasonic welding is applied to bond the wire. Quality suffers if the IC and the frame are not well connected. The

Table 4.9: Pull strength in grams for 10 units with 5 wires each

| wire | Un1 | Un2 | Un3 | Un4 | Un5 | Un6 | Un7 | Un8 | Un9 | Un10 |
|------|------|------|------|------|------|------|------|------|------|------|
| 1 | 11.6 | 9.7 | 11.3 | 10.1 | 10.9 | 9.7 | 9.6 | 10.7 | 10.8 | 9.9 |
| 2 | 11.3 | 11.2 | 11.0 | 10.2 | 10.8 | 11.0 | 10.3 | 11.2 | 10.6 | 9.4 |
| 3 | 10.3 | 9.8 | 10.2 | 10.8 | 10.3 | 10.3 | 9.2 | 9.9 | 9.2 | 11.3 |
| 4 | 11.7 | 11.0 | 11.1 | 8.6 | 11.7 | 10.8 | 9.1 | 9.7 | 10.6 | 10.8 |
| 5 | 10.3 | 9.6 | 11.4 | 10.8 | 10.6 | 9.3 | 10.2 | 9.7 | 10.5 | 11.3 |

bond pull test is used to pull on the wire to determine the strength required to disconnect the wire from the lead frame.

Two experiments were designed to establish the pull strength repeatability $\sigma_e^2$. In the first experiment, the $\sigma_e^2$ was confounded with $\sigma_w^2$, the variability created by the wire position on the production unit. In the second experiment, $\sigma_e^2$ was confounded with the variability $\sigma_u^2$ caused by different production units. The information combined from two experiments enables separate estimation of $\sigma_e{}^2$, $\sigma_w^2$, $\sigma_u^2$, as well as of the variability $\sigma_{oper}^2$ due to different operators.

**Experiment 1: Description and Data**

Ten units were selected randomly from a batch of units produced under identical conditions. Five wire positions were selected randomly and one operator carried out the pull test for each wire. (The wire positions were the same for all units). The experiment results are presented in Table 4.9.[1]

This experiment in fact has an hierarchical design; see Fig. 4.2. The units play the role of batches, and the wires play the role of the samples within a batch. The only difference is that there are no repeated measurements of the same sample. In the notation of Sect. 4.3, $I = 10, J = 5$ and $K = 1$.

The random pull strength $X_{ij}$ of the $j$th wire in the $i$th unit is modeled as

$$X_{ij} = \mu + A_i + B_{j(i)} + \epsilon_{ij}. \tag{4.5.1}$$

Here $\mu$ is the overall mean value. $A_i$ is the random contribution due to the variability between units, and is a zero-mean random variable with variance $\sigma_u^2$. $B_{j(i)}$ is the random contribution due to the wire position within a unit, and is a zero-mean random variable with variance $\sigma_w^2$. $\epsilon_{ij}$ is the random contribution of the pull strength for fixed unit and fixed wire position. It is assumed that the $\epsilon_{ij}$ are zero mean random variables with variance $\sigma_e^2$.

---

[1]Source: Teresa Mitchell, Victor Hegemann and K.C. Liu "GRP methodology for destructive testing and quantitative assessment of gauge capability", pp. 47-59, in the collection by Veronica Czitrom and Patrick D. Spagon *Statistical Case Studies for Industrial Process Improvement* ©1997. Borrowed with the kind permission of the ASA and SIAM.

### Experiment 1: Calculations

Let us modify appropriately the formulas (4.3.5)–(4.3.7) for our data. Obviously, $ss_e = 0$. Also

$$ss_{B(A)} = \sum_{i=1}^{10} \sum_{j=1}^{5} (x_{ij} - \bar{x}_{i\cdot})^2, \qquad (4.5.2)$$

where $x_{ij}$ is the observed pull strength for unit $i$ and wire $j$, and $\bar{x}_{i\cdot}$ is the average pull strength for unit $i$. Finally,

$$ss_A = 5 \cdot \sum_{i=1}^{10} (\bar{x}_{i\cdot} - \bar{x}_{\cdot\cdot})^2, \qquad (4.5.3)$$

where $\bar{x}_{\cdot\cdot}$ is the overall average pull strength.

Now we have the following two equations (compare with Table 4.5 and set $\sigma_A = \sigma_u, \sigma_B = \sigma_w$):

$$\frac{ss_A}{9} = \hat{\sigma}_e^2 + \hat{\sigma}_w^2 + 5\hat{\sigma}_u^2; \qquad (4.5.4)$$

$$\frac{ss_{B(A)}}{40} = \hat{\sigma}_e^2 + \hat{\sigma}_w^2. \qquad (4.5.5)$$

It is easy to find the values of $ss_A$ and $ss_{B(A)}$, for example by using the one-way ANOVA procedure in Statistix. The results are below:

$$ss_A = 8.12, \; ss_{B(A)} = 19.268.$$

From (4.5.4) and (4.5.5), we find that

$$\hat{\sigma}_e^2 + \hat{\sigma}_w^2 = 0.482; \qquad (4.5.6)$$
$$\hat{\sigma}_u^2 = 0.084. \qquad (4.5.7)$$

### Experiment 2: Description and Data

Thirty units were chosen randomly from the same production lot as in Experiment 1. Three randomly chosen operators carried out the pull strength test. Each operator pulled one wire on each of ten units. The wires subjected to the pulling test had the same positions on each unit. The pull strength measurement results are presented in Table 4.10.[2]

Experiment 2 also has a hierarchical design. Operators play the role of batches, and the units are the samples within batch. Because of the nature of the experiment, repeated measurements for the same sample are not possible.

---

[2] Source: Teresa Mitchell, Victor Hegemann and K.C. Liu "GRP methodology for destructive testing and quantitative assessment of gauge capability", p. 52, in the collection by Veronica Czitrom and Patrick D. Spagon *Statistical Case Studies for Industrial Process Improvement* ©1997. Borrowed with the kind permission of the ASA and SIAM.

Table 4.10: Pull strength in grams for one wire on each of ten units

| Unit | Operator 1 | Operator 2 | Operator 3 |
|------|-----------|-----------|-----------|
| 1 | 10.3 | 9.3 | 11.1 |
| 2 | 10.9 | 11.4 | 10.4 |
| 3 | 11.0 | 10.1 | 11.5 |
| 4 | 11.7 | 9.1 | 11.1 |
| 5 | 9.8 | 10.0 | 11.6 |
| 6 | 10.5 | 9.9 | 9.7 |
| 7 | 10.5 | 11.1 | 10.9 |
| 8 | 10.8 | 10.9 | 9.8 |
| 9 | 10.8 | 10.3 | 11.4 |
| 10 | 10.3 | 9.9 | 10.4 |

In this experiment $I = 3$, $J = 10$ and $K = 1$. Denote by $y_{ij}$ the measured pull strength recorded by operator $i$ and unit $j$. For each $i$, $j$ ranges from 1 to 10:

$$y_{ij} = \mu^\star + C_i + A^\star_{j(i)} + \epsilon^\star_{ij}, \qquad (4.5.8)$$

where $\mu^\star$ is the overall mean, $C_i$ is the variability due to the operator, $A^\star_{j(i)}$ is the variability due to the unit for a fixed operator, and $\epsilon^\star_{ij}$ is the random repeatability contribution to the pull strength for a fixed wire, operator and unit. (Imagine that there exist several absolutely identical copies of the wire bond on a fixed position for each $i$ and $j$. Each such bond would produce an independent replica of the random variable $\epsilon^\star_{ij}$). Our assumptions are that all random variables involved have zero mean value and the following variances:

$\text{Var}[C_i] = \sigma^2_{oper}$;
$\text{Var}[A_{j(i)}] = \sigma^2_u$, equal to the corresponding variance in the Experiment 1;
$\text{Var}[\epsilon^\star_{ij}] = \sigma^2_e$, equal to the variance of $\epsilon_{ij}$ in Experiment 1.

It is important to note that all units in both experiments are produced in identical conditions, so that the tool repeatability expressed as $\sigma^2_e$ is the same in both experiments.

Similarly to Experiment 1, we will use the following two sum of squares:

$$ss_{A(C)} = \sum_{i=1}^{3}\sum_{j=1}^{10}(y_{ij} - \bar{y}_{i.})^2; \qquad (4.5.9)$$

$$ss_C = 10\sum_{i=1}^{3}(\bar{y}_{i.} - \bar{y}_{..})^2. \qquad (4.5.10)$$

Here $\bar{y}_{i.}$ is the mean of all measurements for operator $i$, and $\bar{y}_{..}$ is the mean of all 30 observations. Now we have the following two equations (compare with Table 4.5):

$$\frac{ss_C}{2} = \hat{\sigma}_e^2 + \hat{\sigma}_u^2 + 10\hat{\sigma}_{oper}^2; \tag{4.5.11}$$

$$\frac{ss_{A(C)}}{27} = \hat{\sigma}_e^2 + \hat{\sigma}_u^2. \tag{4.5.12}$$

### Calculations Completed

It is easy to compute, using single-factor ANOVA, that $ss_{A(C)} = 11.55$ and $ss_C = 1.92$. Substituting into (4.5.11) and (4.5.12), we obtain that

$$\hat{\sigma}_e^2 + \hat{\sigma}_u^2 = 0.43, \tag{4.5.13}$$

$$\hat{\sigma}_{oper}^2 = 0.053. \tag{4.5.14}$$

Combining this with the result of Experiment 1, we obtain the following estimates of all variances and standard deviations appearing in the model:

$\hat{\sigma}_e^2 = 0.346; \ \hat{\sigma}_e = 0.59;$
$\hat{\sigma}_u^2 = 0.084; \ \hat{\sigma}_u = 0.29;$
$\hat{\sigma}_{oper}^2 = 0.053; \ \hat{\sigma}_{oper} = 0.23;$
$\hat{\sigma}_w^2 = 0.136; \ \hat{\sigma}_w = 0.37.$

Mitchell et al. (1997) were interested in estimating the parameter $\sigma_{meas} = \sqrt{\sigma_e^2 + \sigma_{oper}^2}$. This estimate is $\hat{\sigma}_{meas} = \sqrt{0.346 + 0.053} = 0.63$.

## 4.6   Complements and Exercises

**1. Range method for estimating repeatability**
The following simple procedure produces good point estimates of $\sigma_e$ in the R&R studies; see Montgomery and Runger (1993). Compute, for each item $i$ and each operator $j$ the range $r_{i,j}$ for $K$ measurements in the $(i,j)$th cell; see Table 4.7. For example, for $i = 2, j = 1 \ r_{2,1} = 0.004$. Compute the mean range over all $I \times J$ cells:

$$\overline{R} = \frac{\sum_{i=1}^{I} \sum_{j=1}^{J} r_{i,j}}{IJ}. \tag{4.6.1}$$

Estimate $\sigma_e$ as

$$\tilde{\sigma}_e^\star = \frac{\overline{R}}{A_K}, \tag{4.6.2}$$

Table 4.11: Running times of 20 mechanical time fuses measured by operators stopping two independent clocks (Kotz and Johnson 1983, p. 542)

| Fuse No | First instrument (sec) | Second instrument (sec) |
|---------|------------------------|-------------------------|
| 1 | 4.85 | 5.09 |
| 2 | 4.93 | 5.04 |
| 3 | 4.75 | 4.95 |
| 4 | 4.77 | 5.02 |
| 5 | 4.67 | 4.90 |
| 6 | 4.87 | 5.05 |
| 7 | 4.67 | 4.90 |
| 8 | 4.94 | 5.15 |
| 9 | 4.85 | 5.08 |
| 10 | 4.75 | 4.98 |
| 11 | 4.83 | 5.04 |
| 12 | 4.92 | 5.12 |
| 13 | 4.74 | 4.95 |
| 14 | 4.99 | 5.23 |
| 15 | 4.88 | 5.07 |
| 16 | 4.95 | 5.23 |
| 17 | 4.95 | 5.16 |
| 18 | 4.93 | 5.11 |
| 19 | 4.92 | 5.11 |
| 20 | 4.89 | 5.08 |

where $A_K$ is the constant from Table 2.6. $K$ is the number of measurements made by one operator. In the notation of Table 2.6, $K = n = 3$. In our case, we must take $A_3 = 1.693$.

**2.** *Grubbs's method of analysing measurement data when repeated measurements are not available*

The data in Table 4.11 give the burning time in seconds of 20 shell fuses. For each shell, the burning time was measured by two independent instruments. For obvious reasons, these are *not* repeated measurements.[1]

Let us formulate the model corresponding to this measurement scheme. Denote by $Y_{i1}$ the first instrument reading time for fuse $i$. We assume that

$$Y_{i1} = X_i + \beta_1 + \epsilon_{i1}, \tag{4.6.3}$$

[1] Borrowed from S. Kotz and N.L. Johnson (eds.) *Encyclopedia of Statistical Sciences*, Vol.3, p. 543; copyright ©1983 John Wiley & Sons, Inc. This material is used by permission of John Wiley & Sons, Inc.

where $X_i$ is the true burning time of the $i$th fuse, $\beta_1$ is an unknown constant (instrument 1 bias), and $\epsilon_{i1}$ is the random error of instrument 1. It is assumed that $X_i$ and $\epsilon_{i1}$ are independent random variables,

$$X_i \sim N(\mu, \sigma_X^2), \quad \epsilon_{i1} \sim N(0, \sigma_{e1}^2). \tag{4.6.4}$$

Similarly, the reading of the second instrument for the $i$th fuse is

$$Y_{i2} = X_i + \beta_2 + \epsilon_{i2}, \tag{4.6.5}$$

where $\beta_2$ is an unknown constant bias of instrument 2, and $\epsilon_{i2}$ is the random measurement error of instrument 2. It is assumed that $X_i$ and $\epsilon_{i2}$ are independent,

$$\epsilon_{i2} \sim N(0, \sigma_{e2}^2). \tag{4.6.6}$$

Find estimates of $|\beta_1 - \beta_2|$, $\sigma_X^2$, $\sigma_{e1}^2$ and $\sigma_{e2}^2$.

It may seem that there are not enough data for this estimation. The trick is to involve two additional statistics – the sum and the difference of the readings of the instruments.

*Solution*

Denote $SUM = Y_{i1} + Y_{i2}$, $DIFF = Y_{i1} - Y_{i2}$. It is easy to prove that

$$\text{Var}[Y_{i1}] = \sigma_X^2 + \sigma_{e1}^2, \tag{4.6.7}$$

$$\text{Var}[Y_{i2}] = \sigma_X^2 + \sigma_{e2}^2, \tag{4.6.8}$$

$$\text{Var}[SUM] = 4\sigma_X^2 + \sigma_{e1}^2 + \sigma_{e2}^2, \tag{4.6.9}$$

$$\text{Var}[DIFF] = \sigma_{e1}^2 + \sigma_{e2}^2. \tag{4.6.10}$$

It follows from these formulas that

$$\sigma_X^2 = \frac{\text{Var}[SUM] - \text{Var}[DIFF]}{4}, \tag{4.6.11}$$

$$\sigma_{e1}^2 = \text{Var}[Y_{i1}] - \sigma_X^2, \tag{4.6.12}$$

and

$$\sigma_{e2}^2 = Var[Y_{i2}] - \sigma_X^2, \tag{4.6.13}$$

$$E[Y_{i1}] - E[Y_{i2}] = \beta_1 - \beta_2. \tag{4.6.14}$$

**3.** For the data in Table 4.11, estimate $\sigma_X, \sigma_{e1}, \sigma_{e2}$ and $\delta = \beta_1 - \beta_2$.

A good source on Grubbs' estimation method are the papers by Grubbs (1948; 1973); see also Kotz and Johnson (1983, p. 542). For more than two instruments a separate treatment is needed, but the principle of using pairwise differences and sums remains the same.

**4.** For the data of Table 4.11, construct the 95% confidence interval for the parameter $\delta$.
*Hint:* Use the statistics of the running time differences.

**5.** Compute the variance of $SS_e$ in Example 4.4.1.
*Solution.* According to (4.4.26), $SS_e/\sigma_e^2$ has a chi-square distribution with $\nu = IJ(K-1)$. Recall that if $Y$ has a chi-square distribution with $\nu$ degrees of freedom, then $\mathrm{Var}[y] = 2\nu$. Therefore,

$$\mathrm{Var}[SS_e/\sigma_e^2] = 2\nu,$$

and

$$\mathrm{Var}[SS_e] = 2IJ(K-1)\sigma_e^4. \tag{4.6.15}$$

**6.** Compute the variances of $SS_A, SS_B$ and $SS_{AB}$ in Example 4.4.1.

*Answer:*

$$\mathrm{Var}[SS_A] = 2(I-1)\big(\sigma_e^2 + K\sigma_{AB}^2 + JK\sigma_A^2\big)^2; \tag{4.6.16}$$

$$\mathrm{Var}[SS_B] = 2(J-1)\big(\sigma_e^2 + K\sigma_{AB}^2 + IK\sigma_B^2\big)^2; \tag{4.6.17}$$

$$\mathrm{Var}[SS_{AB}] = 2(I-1)(J-1)\big(\sigma_e^2 + K\sigma_{AB}^2\big)^2. \tag{4.6.18}$$

**7.** For Example 4.3.1, check the null hypotheses for $\sigma_A^2$ and $\sigma_B^2$ at significance level 0.05.

# Chapter 5

# Measurement Uncertainty: Error Propagation Formula

*If a man will begin with certainty, he shall end in doubts; but if he will be content with doubts, he shall end in certainties.*

Francis Bacon, *The Advancement of Learning*

## 5.1 Introduction

So far we have dealt with various aspects of uncertainty in measuring a *single* quantity. In Sect. 4.4 it was a measurement of weight; in Sect. 4.5 we analyzed results from measuring pull strength. In most real-life situations, the measurement process involves *several* quantities whose measurement result is subject to uncertainty. To clarify the exposition, let us consider a rather simple example.

*Example 5.1.1: Measuring specific weight.*
Suppose that there is a need to know the specific weight of a certain steel used in metallic constructions. For this purpose, a cylindrical specimen is prepared from the steel, and the weight and volume of this specimen are measured. Denote by $D$ the diameter of the specimen and by $L$ its length. Then the volume of the specimen is $V = \pi D^2 L/4$. If the weight of the specimen is $W$, then the specific weight $\rho$ equals

$$\rho = \frac{W}{V} = \frac{4W}{\pi D^2 L}.$$ (5.1.1)

Note that all quantities in this formula $W, D, L$ are obtained as the result of a measurement, i.e. they are subject to random errors. To put it more formally, they are random variables. Therefore the specific weight $\rho$ is also a random variable. We are interested in the *variance* of $\rho$, $\text{Var}[\rho]$, or in its standard deviation $\sigma_\rho = \sqrt{\text{Var}[\rho]}$.

An approximate value of $\sigma_\rho$ can be found by means of a formula known as the *error propagation formula* which is widely used in measurements. Let us first derive it, and then apply to our example.

## 5.2  Error Propagation Formula

Let a random variable $Y$ be expressed as a known function $f(\cdot)$ of several other *independent* random variables $X_1, X_2, \ldots, X_n$:

$$Y = f(X_1, X_2, \ldots, X_n). \tag{5.2.1}$$

Let $\mu_i$ and $\sigma_i$ be the mean and standard deviation, respectively, of $X_i$. To obtain an approximate expression for $\text{Var}[Y]$ we will use an approach known as the delta method; see Taylor (1997, p. 146).

Let us expand $Y$ as a Taylor series around the mean values $\mu_i$ of $X_i$, $i = 1, \ldots, n$:

$$Y = f(\mu_1, \mu_2, \ldots, \mu_n) + \sum_{i=1}^{n} \frac{\partial f}{\partial X_i}\bigg|_{X_i = \mu_i} \cdot (X - \mu_i) + \ldots \tag{5.2.2}$$

It this formula the dots represent the higher-order terms. The partial derivatives are evaluated at the mean values of the respective random variables.

Now ignore all higher-order terms and write:

$$Y \approx f(\mu_1, \mu_2, \ldots, \mu_n) + \sum_{i=1}^{n} \frac{\partial f}{\partial X_i}\bigg|_{X_i = \mu_i} \cdot (X_i - \mu_i) \tag{5.2.3}$$

Let us compute the variance of the right-hand side of (5.2.3). Denote the partial derivatives by $f_i'$. Since the terms in the sum are independent, we can use formula (2.1.22):

$$Var[Y] \approx \sum_{i=1}^{n} (f_i')^2 \text{Var}[X_i]. \tag{5.2.4}$$

From this it follows that

$$\sigma_Y \approx \sqrt{\sum_{i=1}^{n} (f_i')^2 \text{Var}[X_i]}. \tag{5.2.5}$$

Expression (5.2.3) implies that

$$\mu_Y = E[Y] \approx f(\mu_1, \ldots, \mu_n). \tag{5.2.6}$$

We are already familiar with the notion of the *coefficient of variation* of a positive random variable $Y$, defined as $\sigma_Y/\mu_Y$; see (2.1.29). Assuming that $Y$ is positive, in measurement practice we often use the term *relative standard deviation* (RSD) instead of coefficient of variation. Let us write the formula for the RSD of $Y$:

$$\frac{\sigma_y}{\mu_Y} \approx \sqrt{\sum_{i=1}^{n}(f_i')^2\left(\frac{\sigma_i}{\mu_Y}\right)^2}. \tag{5.2.7}$$

The use of (5.2.5) and (5.2.7) is a valid operation if the RSD's of random variables $X_i$, $\sigma_i/\mu_i$ are small, say no greater than 0.05. In other words, it will be assumed that

$$\frac{\sigma_i}{\mu_i} \le 0.05. \tag{5.2.8}$$

In measurements, this is usually a valid assumption.

In measurement practice, we do not know, however, the mean values of random variables $Y, X_1, \ldots, X_n$. What we do know are their *observed* values, which we denote as $y$ (or $f$), and $x_1, \ldots, x_n$. In the case where the relative standard deviations of all random variables are small, it is a valid operation to *replace* the theoretical means $\mu_i$ by $x_i$. Thus, we arrive at the following version of (5.2.7):

$$\frac{\sigma_y}{y} \approx \sqrt{\sum_{i=1}^{n}(\widehat{f_i'})^2\left(\frac{\sigma_i}{y}\right)^2}, \tag{5.2.9}$$

where $\widehat{f_i'}$ is the the partial derivative of $f(\cdot)$ with respect to $X_i$, evaluated at the observed values of $X_i = x_i$, $i = 1, \ldots, n$. (5.2.9) is known as the *error propagation formula* (EPF). It may also be writen as

Its equivalent version is

$$\sigma_y \approx \sqrt{\sum_{i=1}^{n}(\widehat{f_i'})^2\sigma_i^2}. \tag{5.2.10}$$

In practice, we do not usually know the variances $\sigma_i^2$ either, and use their estimates $\hat{\sigma}_i^2$. Then the previous formula takes the form

$$\sigma_y \approx \sqrt{\sum_{i=1}^{n}(\widehat{f_i'})^2\hat{\sigma}_i^2}. \tag{5.2.11}$$

*Example 5.1.1 continued.*
The following measurement results of $W, D$ and $L$ were obtained: $W = 25.97$ gram; $D = 1.012$ cm; $L = 4.005$ cm. Assume that it is known from previous

experience that the standard deviations in measuring the weight $W$ and the
quantities $D, L$ are as follows:

$$\sigma_W = 0.05 \text{ g;} \sigma_D = 0.002 \text{ cm; } \sigma_L = 0.004 \text{ cm.}$$

Find the value of $\rho$ and its standard deviation $\sigma_\rho$ .
*Solution*
Here $\rho$ plays the role of $Y$, and $W, L, D$ play the role of $X_1, X_2, X_3$. The function
$f(W, L, D)$ is

$$\rho = 4W/(\pi D^2 L). \tag{5.2.12}$$

The estimated value of $\rho$ is $\hat{\rho} = 4 \cdot 25.97/(\pi \cdot 1.012^2 \cdot 4.005) = 8.062 \text{ g/cm}^3$.
   Compute the partial derivatives $\partial\rho/\partial w, \partial\rho/\partial L, \partial\rho/\partial D$ and evaluate the partial derivatives at the observed values of $W, D$ and $L$:

$$f'_W = \frac{\partial\rho}{\partial W} = \frac{4}{\pi D^2 L}; \tag{5.2.13}$$

$$f'_L = \frac{\partial\rho}{\partial L} = \frac{-4W}{\pi D^2 L^2}; \tag{5.2.14}$$

$$f'_D = \frac{\partial\rho}{\partial D} = \frac{-8W}{\pi L D^3}.$$

Substituting into these formulas the observed values of $W, D, L$, we obtain:

$$(\hat{f}_W)^2 = 0.096; \tag{5.2.15}$$

$$(\hat{f}_L)^2 = 4.052; \tag{5.2.16}$$

$$(\hat{f}_D)^2 = 253.8.$$

By (5.2.11) the standard deviation of $\rho$ is

$$\sigma_\rho \approx \sqrt{0.096 \cdot 0.05^2 + 4.052 \cdot 0.004^2 + 253.8 \cdot 0.002^2} = 0.036. \tag{5.2.17}$$

It is a common practice to present the computation result in the form $\hat{\rho} \pm \sigma_\rho$ :
$8.062 \pm 0.036 \text{ gram/cm}^3$.

## 5.3  EPF for Particular Cases of $Y = f(X_1, \ldots, X_n)$

**Case 1. EPF for $Y = \sum_{i=1}^{n} c_i X_i$**
In this case $f$ is a linear function of its arguments. The Taylor series expansion
produces the same result as formula (2.1.22):

$$\sigma_Y = \sqrt{\text{Var}[Y]} = \sqrt{\sum_{i=1}^{n} c_i^2 \sigma_i^2}, \tag{5.3.1}$$

where $\sigma_i^2 = \text{Var}[X_i]$. (Recall that the $X_i$ are assumed to be independent random variables.) If the $\sigma_i^2$ are replaced by their estimates $\hat{\sigma}_i^2$, the analogue of (5.2.11) is

$$\sigma_Y = \sqrt{\text{Var}[Y]} \approx \sqrt{\sum_{i=1}^{n} c_i^2 \hat{\sigma}_i^2}, \tag{5.3.2}$$

**Case 2. EPF for $Y = f(X_1, X_2) = X_1/(X_1 + X_2)$**

In this case,

$$f_1' = X_2/(X_1 + X_2)^2; f_2' = -f/(X_1 + X_2). \tag{5.3.3}$$

If we observe $X_1 = x_1, X_2 = x_2$, and $\hat{\sigma}_i^2$ is the estimate of $\sigma_i^2$, $i = 1, 2$, and $\hat{f} = x_1/(x_1 + x_2)$, then (5.2.11) takes the form

$$\sigma_y \approx \frac{\sqrt{\hat{\sigma}_1^2 \cdot x_2^2 + \hat{\sigma}_2^2 \cdot x_1^2}}{(x_1 + x_2)^2}. \tag{5.3.4}$$

**Case 3. EPF for $Y = f(X_1, \ldots, X_k, X_{k+1}, \ldots, X_n) = \prod_{i=1}^{k} X_i / \prod_{i=k+1}^{n} X_i$**

Very often the function $f(\cdot)$ is a ratio of products of random variables. For this particular case, one can check that

$$\left(f_i'\right)^2 = \frac{f^2}{X_i^2}, i = 1, \ldots, n.$$

Replace the $X_i$ by their observed values $x_i$ and denote $f(x_1, \ldots, x_n) = \hat{f}$. Now formula (5.2.11) takes the following elegant form:

$$\sigma_y \approx |\hat{f}| \sqrt{\sum_{i=1}^{n} \left(\frac{\sigma_i}{x_i}\right)^2}. \tag{5.3.5}$$

Another form of this formula is

$$\frac{\sigma_y}{|\hat{f}|} \approx \sqrt{\sum_{i=1}^{n} \left(\frac{\sigma_i}{x_i}\right)^2}. \tag{5.3.6}$$

In words: the relative standard deviation (RSD) of the resulting variable $Y$ is approximately equal to the square root of sum of squares of the RDS's of the variables $X_1, \ldots, X_n$.

**4. EPF for $Y = f(X_1, X_2, X_3) = \sqrt{X_1/C_1 + X_2/C_2 + X_3/C_3}$**

In this case,

$$f_i' = \frac{C_i^{-1}}{2Y}, \tag{5.3.7}$$

$i = 1, 2, 3$. Put $\sigma_i^2 = \mathrm{Var}[X_i]$. Then we obtain that

$$\sigma_Y \approx \frac{\sqrt{\sigma_1^2/C_1^2 + \sigma_2^2/C_2^2 + \sigma_3^2/C_3^2}}{2|Y|}. \tag{5.3.8}$$

In practice, we must replace $Y$ by its estimate (its observed value) $\widehat{Y}$, and $\sigma_i^2$ by their estimates $\hat{\sigma}_i^2$. Then

$$\sigma_Y \approx \frac{\sqrt{\hat{\sigma}_1^2/C_1^2 + \hat{\sigma}_2^2/C_2^2 + \hat{\sigma}_3^2/C_3^2}}{2|\widehat{Y}|}. \tag{5.3.9}$$

We will apply this formula for approximate computation of the standard deviation of $\sigma_{R\&R}$ defined by (4.4.18); see Exercise 4 in the next section.

*Remarks*

1. How do we obtain the relative standard deviations of the quantities entering the formula for $\sigma_Y$? Generally, there are three ways: statistical experiment, i.e. previous repeatability/reproducibility studies; certified standards data for the measuring instrument and/or from manufacturer warranties; expert opinion/analyst judgments.

Useful sources on the methodology and practice of establishing uncertainty are Eurachem (2000) and Kragten (1994).

Sometimes, the certificate data say that the measurement error of a certain device has a specific distribution, for example a uniform one in an interval of given length $\Delta$. Then the corresponding standard deviation is equal to $\sigma = \Delta/\sqrt{12}$.

2. It is important to give the following warnings. Often, the EPF *underestimates* the RSD of the final result. This happens because certain factors which in reality influence the result and may increase the uncertainty are omitted from the formula $y = f(x_1, \ldots, x_n)$. So, for example, the result may depend on the variations of the ambient temperature during the experiment, and the temperature is omitted from the input variables on which the output $Y$ depends.

Another reason for underestimation of the uncertainty is underestimation of component variability. Consider, for example, measuring the concentration $c$ of some chemical agent, $c = W/V$, where $W$ is the weight of the material and $V$ is the volume of the solution. When we measure the weight of the material to be dissolved, we may take into consideration only the variability introduced by the weighing process itself and ignore the fact that the agent is not 100% pure. In fact, the content of the agent has random variations, and an additional statistical experiment should be carried out to establish the variability introduced by this factor.

## 5.4 Exercises

**1.** Specific weight is computed as $\rho = W/V$, where $W$ is the weight measurement and $V$ is the volume measurement. The measurement results are $w = 13.505$ g, $v = 3.12$ cm$^3$. The relative standard deviations of the weight and volume measurements are 0.01 and 0.025, respectively. Find the RSD of the specific weight.

*Answer:* By (5.3.6) $\sigma_\rho/\rho \approx \sqrt{0.025^2 + 0.01^2} = 0.027$.

**2.** The volume $V$ of a rectangular prism equals $V_1 V_2 V_3$ where $V_1$ and $V_2$ are the dimensions of its base and $V_3$ is its height. Each of these dimensions is measured with RSD of 0.01. Find the relative error of the volume measurement.

*Answer:* $\sqrt{3 \cdot 0.01^2} = 0.0017$.

**3.** Potassium hydrogen phthalate has the following chemical formula: $C_8H_5O_4K$. Its atomic weight equals $8 \cdot W_C + 5 \cdot W_H + 4 \cdot W_O + W_K$. $W_C, W_H, W_O, W_K$ are the atomic weights of elements C, H, O, K respectively; see Table 2.4. The standard deviation of the estimate of the atomic weight of each element is given in the third column of Table 2.4. For example, $0.00058 = \sqrt{\text{Var}[W_C]}$.
　　Find the approximate standard deviation (standard error) in estimating the atomic weight of potassium hydrogen phthalate.

*Solution:* The relevant expression is (5.3.2).

$$\sigma_W \approx \sqrt{64 \cdot 0.00058^2 + 25 \cdot 0.000040^2 + 16 \cdot 0.00017^2 + 0.000058^2} = 0.0047.$$

**4.** *Estimation of standard error of* $\sigma_{R\&R}$
The estimate of $\sigma_{R\&R}$ is defined by (4.4.18). If we treat $\sigma_{R\&R}$ as a random variable, then the corresponding formula will be the following:

$$\sigma_{R\&R} = \sqrt{\frac{SS_B}{IK(J-1)} + \frac{SS_{AB}}{IK(J-1)} + \frac{SS_e}{IJK}}, \qquad (5.4.10)$$

see Case 4 of Sect. 5.3.
　　Calculate an approximate value of the standard error of $\sigma_{R\&R}$. Take $I = 2, J = 5, K = 3$ and $\hat{\sigma}_{R\&R} = 0.0079$. Use the results of Example 4.4.1.

*Solution.* The main difficulty is obtaining estimates of variances of $SS_e, SS_{AB}$ and $SS_B$. For this purpose, we must use formulas (4.6.15), (4.6.17), (4.6.18). These contain $\sigma_e^2, \sigma_B^2, \sigma_{AB}^2$.

By (4.6.15),

$$\widehat{\text{Var}}[SS_e] = 40 \cdot \hat{\sigma}_e^2 = 40 \cdot (0.0071)^4 = 1.02 \cdot 10^{-7}. \tag{5.4.11}$$

To obtain an estimate of $\sigma_{AB}^2$, use the following formula which is derived from (4.4.16):

$$\hat{\sigma}_{AB}^2 = \max\left(0, \left(ss_{AB}/4 - \hat{\sigma}_e^2\right)/3\right).$$

Substituting into this formula $ss_{AB} = 0.0000176$ and $\hat{\sigma}_e^2 = 0.0071^2$, we see that $\hat{\sigma}_{AB}^2 = 0$. Now by (4.4.15),

$$\hat{\sigma}_B^2 = (ss_B/4 - \hat{\sigma}_e^2)/6 = 1.25 \cdot 10^{-5}.$$

Now by (4.6.17), (4.6.18),

$$\widehat{\text{Var}}[SS_B] = 8(0.0071^2 + 6 \cdot 1.25 \cdot 10^{-5})^2 = 1.26 \cdot 10^{-7}, \text{ and}$$
$$\widehat{\text{Var}}[SS_{AB}] = 8 \cdot 0.0071^4 = 2.03 \cdot 10^{-8}.$$

Now substitute all estimates into the formula

$$\text{St. dev. of } \hat{\sigma}_{R\&R} \approx \frac{\sqrt{\widehat{Var}[SS_B]/24^2 + \widehat{Var}[SS_{AB}]/24^2 + \widehat{Var}[SS_e]/30^2}}{2\hat{\sigma}_{R\&R}}.$$

The final result is 0.0012 and this is the approximate value of the standard deviation (standard error) of $\hat{\sigma}_{R\&R}$.

The above calculations are an implementation of the method suggested by Vardeman and VanValkenburg (1999) based on the use of EPF.

# Chapter 6

# Calibration of Measurement Instruments

*A little inaccuracy sometimes saves tons of explanation.*

Saki

## 6.1 Calibration Curves

### 6.1.1 Formulation of the Problem

The best way to formulate the problem of calibration is to consider a typical example. Suppose that we are interested in measuring glucose concentration in mg/dl in certain substances. For this purpose, a spectrophotometric method is used. Without going into detail, let us note that the response of the measurement device $y$ is the so-called *absorbance*, which depends on the concentration $x$. So, the measurement instrument provides us with a value $y$, the response, when the substance has glucose concentration $x$. One might say that each measurement instrument is characterized by its own relationship between $x$ and $y$. Formally, it seems that we can write

$$y = f(x), \tag{6.1.1}$$

where the function $f(\cdot)$ represents this relationship.

It is important to note that what happens in measurements is *not* a purely deterministic relationship like (6.1.1). Suppose that we prepare ten portions of a substance which contains *exactly* 50 mg/dl of glucose. Then we measure the response of our instrument to these ten specimens of the substance and we observe *different* values of the absorbance. Why does this happen? There might be many reasons: measurement errors, small variations in the environmental conditions which influence the absorbance, operator error, etc. So, in fact, the response at concentration $x$ is a random variable $Y$. Assuming that $x$ is known without error, we represent the mathematical model of this situation as follows:

$$Y = f(x) + \epsilon, \tag{6.1.2}$$

where $\epsilon$ is a zero-mean random variable which describes the deviation of the actual response $Y$ from its "theoretical" value $f(x)$. Then, the mean value of the "output" $Y$, given the "input" $x$, equals $f(x)$. Formally,

$$E[Y] = f(x). \tag{6.1.3}$$

Suppose that we know the relationship $y = f(x)$. This is called the *calibration curve*. We are given some substance with *unknown* concentration of glucose. We observe the instrument response $Y^\star$, i.e. we observe a single value of the absorbance. Our task is to determine the true content $x^\star$. Figure 6.1 illustrates this typical situation.

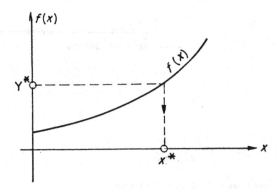

Figure 6.1. Example of a calibration curve

Assume for a moment that $Y = f(x)$ is a totally deterministic and known relationship. Then the problem looks simple: $x^\star$ will be the solution of the equation $Y^\star = f(x^\star)$. Mathematicians would write this via the inverse function:

$$x^\star = f^{-1}(Y^\star). \tag{6.1.4}$$

In real life, our situation is complicated by the fact that we do not know the exact calibration curve $f(x)$. The best we can achieve is a more or less accurate *estimate* of it. Let us use the notation $\widehat{f}(\cdot)$ for this estimate. Thus, we will be solving the equation

$$Y^\star = \widehat{f}(x^\star) \tag{6.1.5}$$

with respect to $x^\star$, and this solution will involve random errors. In other words, the determination of $x^\star$ involves *uncertainty*. This uncertainty depends on the uncertainty involved in constructing the calibration curve.

Table 6.1: Absorbance as a function of glucose in serum

| Concentration of glucose in mg/dl, $x$ | Absorbance, $y$ |
|---|---|
| 0 | 0.050 |
| 50 | 0.189 |
| 100 | 0.326 |
| 150 | 0.467 |
| 200 | 0.605 |
| 400 | 1.156 |
| 600 | 1.704 |

Now we are able to formulate our task as follows:

1) Using the data on the glucose concentration $x$ and and the respective absorbances $Y(x)$, construct the calibration curve $\widehat{f}(x)$.

2) Estimate the uncertainty in solving the inverse problem, i.e. estimate the uncertainty in determining $x^\star$ for given $Y^\star = y^\star$.

## 6.1.2 Linear Calibration Curve, Equal Response Variances

*Example 6.1.1: Measuring glucose concentration*[1]

Mandel (1991, p. 74) reports an experiment set up to construct a calibration curve for an instrument designed to estimate glucose concentration in serum. For this purpose, a series of samples with exactly known glucose concentration were analyzed on the instrument, and the so-called absorbance was measured for each sample. The results are presented in Table 6.1.

The data are a set of pairs $[x_i, y_i], i = 1, 2, \ldots, n$, where $x_i$ are known quantities, i.e. nonrandom variables, and $y_i$ are the observed values of random variables, the responses.

Our principal assumption is that there exists the following relationship between the concentration $x$ and the response (absorbance) $Y$:

$$Y = \alpha + \beta x + \epsilon. \tag{6.1.6}$$

Note that we use capital $Y$ for the random response which depends on $x$ and is influenced also by random measurement error $\epsilon$. For the $i$th measurement, the relationship (6.1.6) can be rewritten as

$$Y_i = \alpha + \beta x_i + \epsilon_i. \tag{6.1.7}$$

---

[1] Reprinted from John Mandel, *Evaluation and Control of Measurements* (1991), by courtesy of Marcel Dekker, Inc.

We assume that $\epsilon_i \sim N(0, \sigma^2)$, and that the random variables $\epsilon_i$ are independent for $i = 1, \ldots, n$.

It follows from (6.1.7) that the mean response $E[Y_i]$ at point $x_i$ equals $f(x_i) = \alpha + \beta x_i$. This is our "ideal" calibration curve. It is assumed therefore that $f(x)$ is a straight line.

Another point of principal importance is that the variance of measurement errors $\sigma^2$ *does not* depend on $x_i$. This is the so-called equal response variance case. We will comment later on how a calibration curve might be constructed when error variances depend on the input variable values $x$. In further exposition we denote by $y_i$ the *observed* value of $Y_i$.

## Constructing the Calibration Curve

For our model (6.1.7), we have to find the values of two parameters determining the linear relationship, $\alpha$ (the intercept) and $\beta$ (the slope). These are found using the so-called least squares principle. According to this, we take estimates of $\alpha$ and $\beta$ as those values $\widehat{\alpha}$ and $\widehat{\beta}$ which *minimize* the following sum of squares:

$$\text{Sum of Squares} = \sum_{i=1}^{n} (y_i - \alpha - \beta x_i)^2. \tag{6.1.8}$$

The geometry of this approach is illustrated in Fig. 6.2.

Finding the estimates of $\alpha$ and $\beta$ is a routine task: compute the partial derivatives of the "Sum of Squares" with respect to $\alpha$ and $\beta$, equate them to zero, and then solve the equations. The details of the solution can be found in any statistics course, we will present the results.

First, define the following quantities, which are all we need to extract from the data:

$$\bar{x} = \sum_{i=1}^{n} x_i/n; \tag{6.1.9}$$

$$\bar{y} = \sum_{i=1}^{n} y_i/n; \tag{6.1.10}$$

$$s_{xx} = \sum_{i=1}^{n} (x_i - \bar{x})^2; \tag{6.1.11}$$

$$s_{yy} = \sum_{i=1}^{n} (y_i - \bar{y})^2; \tag{6.1.12}$$

$$s_{xy} = \sum_{i=1}^{n} (x_i - \bar{x})(y_i - \bar{y}). \tag{6.1.13}$$

Now the *estimates* of $\alpha$ and $\beta$ are:

$$\widehat{\beta} = \frac{s_{xy}}{s_{xx}}, \tag{6.1.14}$$

$$\widehat{\alpha} = \overline{y} - \widehat{\beta}\overline{x}. \tag{6.1.15}$$

In addition, we are able to find an estimate of $\sigma^2$:

$$\widehat{\sigma}^2 = \frac{\sum_{i=1}^n (y_i - \widehat{\alpha} - \widehat{\beta}x_i)^2}{n-2}. \tag{6.1.16}$$

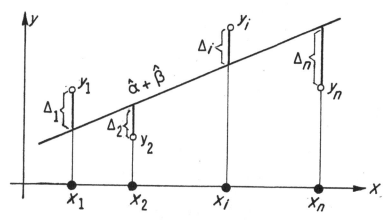

Figure 6.2. The estimates $\widehat{\alpha}$ and $\widehat{\beta}$ minimize $\sum_{i=1}^n \Delta_i^2$.

*Example 6.1.1 continued*
For the data of Example 6.1.1 we have

$\overline{x} = 214.29$; $\overline{y} = 0.6424$; $s_{xx} = 273,570$; $s_{xy} = 754.26$; $s_{yy} = 2.0796$;
$\widehat{\beta} = 0.00276$; $\widehat{\alpha} = 0.0516$; $\widehat{\sigma} = 0.00188$.

Using Statistix, we obtain all the desired results automatically, as shown in Fig. 6.3; see Analytical Software (2000).

### Estimating $x^\star$ from the Observed Response $y^\star$

Suppose we test a certain substance with unknown value of $x$. Our instrument provides us with an observed value of the response $y^\star$. Then the unknown value of the independent variable $x$, $x^\star$, will be obtained by solving the equation

$$y^\star = \widehat{\alpha} + \widehat{\beta}x^\star. \tag{6.1.17}$$

Obviously,

$$x^\star = \frac{y^\star - \widehat{\alpha}}{\widehat{\beta}}. \tag{6.1.18}$$

More convenient will be an equivalent formula:

$$x^\star = \overline{x} + \frac{y^\star - \overline{y}}{\widehat{\beta}},\tag{6.1.19}$$

which one can obtain by substituting (6.1.15) for $\widehat{\alpha}$. It follows from (6.1.19) that

$$(x^\star - \overline{x})^2 = \frac{(y^\star - \overline{y})^2}{\widehat{\beta}^2}.\tag{6.1.20}$$

```
UNWEIGHTED LEAST SQUARES LINEAR REGRESSION OF ABSORB

PREDICTOR
VARIABLES      COEFFICIENT     STD ERROR     STUDENT'S T        P
---------      -----------     ---------     -----------     ------
CONSTANT          0.05163       0.00105          49.28       0.0000
CONCENTR          0.00276     3.593E-06          767.30      0.0000

R-SQUARED              1.0000    RESID. MEAN SQUARE  (MSE)   3.532E-06
ADJUSTED R-SQUARED     1.0000    STANDARD DEVIATION            0.00188

SOURCE         DF       SS           MS           F         P
----------     ---   ----------   ----------    -----    ------
REGRESSION      1      2.07954     2.07954588753.23     0.0000
RESIDUAL        5    1.766E-05   3.532E-06
TOTAL           6      2.07956

CASES INCLUDED 7    MISSING CASES 0
```

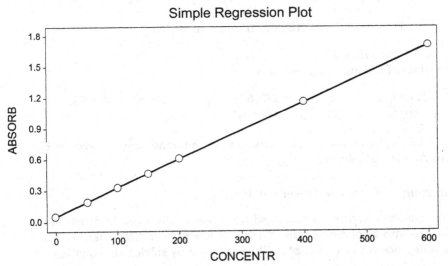

Figure 6.3. The printout and regression plot produced by Statistix.

Suppose that in Example 6.1.1 we observe the absorbance $y^\star = 0.147$. Then by (6.1.19),

$$x^\star = 214.29 + (0.147 - 0.6424)/0.00276 = 34.80.$$

### Uncertainty in Estimating $x^\star$

An important practical issue is estimating the *uncertainty* introduced into the value of $x^\star$ by our measurement procedure. Let us have a closer look at (6.1.19). The right-hand side of it depends on $y^\star$, $\bar{y}$ and $\widehat{\beta}$. Imagine that we repeated the whole experiment, for the same fixed values of $x_1, \ldots, x_n$. Then we would observe values of the response variable $y$ *different* from the previous ones because of random errors $\epsilon_i$. Therefore, we will obtain different values of $\bar{y}$ and different values of the estimate $\widehat{\beta}$. In addition, the response of the measurement instrument to the same unknown quantity of $x$ will also be different because $y^\star$ is subject to random measurement error.

Formally speaking, $x^\star$ is a random variable depending on random variables $\bar{y}, \widehat{\beta}$ and $y^\star$. To keep the notation simple, we use the same small italic letters for these random variables.

Our goal is to obtain an approximate expression for the variance of $x^\star$. In practice, the following approximate formula for the variance of $x^\star$ is used:

$$\mathrm{Var}[x^\star] \approx (x^\star - \bar{x})^2 \left[ \frac{1 + 1/n}{(y^\star - \bar{y})^2} + \frac{1}{\widehat{\beta}^2 s_{xx}} \right] \widehat{\sigma}^2. \tag{6.1.21}$$

(The corresponding expression in Mandel (1991, p. 81) contains a misprint. Expression (6.1.21) coincides with (5.9) in Miller and Miller (1993, p. 113).

Let us present the derivation of this formula. It is an application of the error propagation formula developed in Chap. 5. Although this derivation is instructive, the reader may skip it and simply use the final result. To derive (6.1.21) we need to assume that $\epsilon_i \tilde{N}(0, \sigma^2)$.

### Derivation of (6.1.21)

The following statistical facts are important:
1. $y^\star$ is a random variable with variance $\sigma^2$, the same variance possessed by all observations for various $x$-values.
2. $y^\star$ and $\bar{y}$ are independent random variables since they depend on independent sets of observations.
3. The variance of $\bar{y}$ is $\sigma^2/n$.
4. The variance of $\widehat{\beta}$

$$\mathrm{Var}[\widehat{\beta}] = \frac{\sigma^2}{s_{xx}^2}. \tag{6.1.22}$$

5. Under the normality assumptions, $y^\star - \bar{y}$ and $\widehat{\beta}$ are *independent* random variables.

Properties **4** and **5** are established in statistics courses.

Expression (6.1.19) is of type $Z = C + U/V$, where $U$ and $V$ are independent random variables. Since $C$ is a nonrandom quantity, it does not influence the variance. The expression $U/V$ is an example of case **3** in Sect. 5.3. Using (5.3.5), we obtain that

$$\mathrm{Var}\left[\frac{U}{V}\right] \approx \frac{U^2}{V^2}\left[\frac{\mathrm{Var}[U]}{U^2} + \frac{\mathrm{Var}[V]}{V^2}\right]. \qquad (6.1.23)$$

Put $U = y^\star - \bar{y}$ and $V = \hat{\beta}$. Then $U^2/V^2 = (x^\star - \bar{x})^2$; see (6.1.20). Thus

$$\mathrm{Var}[x^\star] \approx (x^\star - \bar{x})^2 \left[\frac{\mathrm{Var}[y^\star] + \mathrm{Var}[\bar{y}]}{(y^\star - \bar{y})^2} + \frac{\mathrm{Var}[\hat{\beta}]}{\hat{\beta}^2}\right]. \qquad (6.1.24)$$

Replacing the random variables by their observed values, we arrive at the desired formula:

$$\mathrm{Var}[x^\star] \approx (x^\star - \bar{x})^2 \left[\frac{1+1/n}{(y^\star - \bar{y})^2} + \frac{1}{\hat{\beta}^2 s_{xx}}\right] \hat{\sigma}^2. \qquad (6.1.25)$$

*Example 6.1.1 concluded*
Returning to Example 6.1.1, the corresponding numerical result is

$$\mathrm{Var}[x^\star] \approx (34.80 - 214.29)^2 \left(\frac{1+1/7}{(0.147 - 0.6424)^2} + \frac{1}{0.00276^2 \cdot 273,570}\right) \times$$
$$3.532 \cdot 10^{-6} = 0.584,$$

and

$$\sigma^\star = \sqrt{\mathrm{Var}[x^\star]} \approx 0.77.$$

Suppose we think that $\mathrm{Var}[x^\star] = 0.584$ is too large and we would like to carry out a new experiment to reduce it. One way of doing so is to divide the substance with unknown concentration of glucose into several, say $k$, portions and to measure the response for each of these $k$ portions. Then we would observe not a single value of $y^\star$, but a sample of $k$ values $\{y_1^\star, \ldots, y_k^\star\}$. Denote by $\bar{y}^\star$ the corresponding mean value:

$$\bar{y}^\star = \sum_{i=1}^{k} y_i^\star / k. \qquad (6.1.26)$$

Now we will use the value $\bar{y}^\star$ instead of $y^\star$. What will be gained by this? Note that the variance of $\bar{y}^\star$ is $k$ times *smaller* than the variance of $y^\star$. Thus, formula (6.1.21) will take the form

$$\mathrm{Var}[x^\star] \approx (x^\star - \bar{x})^2 \left[\frac{1/k + 1/n}{(y_0 - \bar{y})^2} + \frac{1}{\hat{\beta}^2 s_{xx}}\right] \hat{\sigma}^2. \qquad (6.1.27)$$

For example, let $k = 4$. Then, with the same numerical values for all other variables, we will obtain $\mathrm{Var}[x^\star] \approx 0.237$, $\sigma^\star \approx 0.49$, quite a considerable reduction.

## 6.2 Calibration Curve for Nonconstant Response Variance

### 6.2.1 The Model and the Formulas

In our principal model (6.1.6) it was assumed that the $\epsilon_i$ are zero-mean random variables with *constant* variance. That is to say, it was assumed that the response variance *does not depend* on the value of the input variable $x$. In most technical applications, the variability of the response *does depend* on the value of $x$. For example, for some instruments, the response variability increases with the input variable.

So, let us now assume that,

$$\epsilon_i \sim N(0, \sigma_i^2(x_i)). \tag{6.2.1}$$

The case of nonconstant variances is referred to as *heteroscedasticity*.

How might one discover this phenomenon? A good piece of advice is to use visual analysis of residuals. The residual at point $x_i$ is the difference between the observed value $y_i$ and the corresponding value on the calibration curve $\widehat{\alpha} + \widehat{\beta} x_i$. Formally, the residual $r_i$ is defined as

$$r_i = y_i - (\widehat{\alpha} + \widehat{\beta} x_i). \tag{6.2.2}$$

If the residuals behave, for example, as shown in Fig. 6.4, then obviously the variability of the response is increasing with $x$.

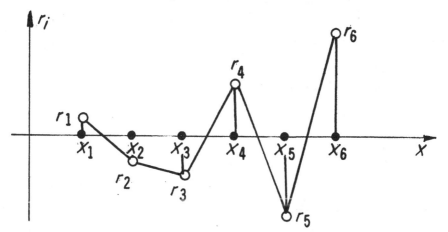

Figure 6.4. The variability of residuals increases with $x$.

We will assume now that the variance of the response at the input value $x_i$ has the following form:

$$\sigma^2(x_i) = \frac{K_v}{w_i}, \tag{6.2.3}$$

where $K_v$ is some positive constant and $w_i$, the so-called *weights*, are assumed to be known exactly or up to a constant multiple. In practice, the weights are either known from previous experience, or on the basis of a specially designed experiment. This experiment must include observing several responses at the $x_i$ values, $i = 1, \ldots, n$.

Transformation is often used to obtain a linear calibration curve. Then it may happen that the transformed responses exhibit heteroscedasticity. The form of the transformation dictates the choice of weights. We will not go into the details of finding the weights. There is an ample literature on this issue; see Madansky (1988, Chapter 2).

One way of obtaining weights is to observe, say, $m$ responses at each $x_i$ and set the weights to be equal to the inverse of the observed sample variances:

$$w_i \approx 1/s_i^2. \tag{6.2.4}$$

So, let us assume that in addition to the data $[x_i, y_i, i = 1, 2, \ldots, n]$ we have a collection of weights

$$w_1, w_2, \ldots, w_n. \tag{6.2.5}$$

In the statistical literature, the procedure of finding the calibration curve when $\text{Var}[\epsilon_i]$ is not constant is called *weighted regression*. The estimates of parameters $\alpha$ and $\beta$ are found by minimizing the following *weighted sum of squares*:

$$\text{Weighted Sum of Squares} = \sum_{i=1}^{n} w_i(y_i - \alpha - \beta x_i)^2. \tag{6.2.6}$$

We see therefore that the squared deviation of $\alpha + \beta x_i$ from $y_i$ is "weighted" in inverse proportion to the variance of $y_i$. Thus, less accurate observations get smaller weight.

Now define the following quantities:

$$\bar{x} = \frac{\sum_{i=1}^{n} w_i x_i}{\sum_{i=1}^{n} w_i}; \tag{6.2.7}$$

$$\bar{y} = \frac{\sum_{i=1}^{n} w_i y_i}{\sum_{i=1}^{n} w_i}; \tag{6.2.8}$$

$$s_{xx} = \sum_{i=1}^{n} w_i(x_i - \bar{x})^2; \tag{6.2.9}$$

$$s_{yy} = \sum_{i=1}^{n} w_i(y_i - \bar{y})^2; \tag{6.2.10}$$

$$s_{xy} = \sum_{i=1}^{n} w_i(x_i - \bar{x})(y_i - \bar{y}). \tag{6.2.11}$$

Table 6.2: Time for water to boil as a function of the amount of water

| $i$ | $x_i$ cm$^3$ | $t_i$ sec | Estimated variance $s_i^2$ | Weight $w_i$ |
|---|---|---|---|---|
| 1 | 100 | 35 | 12.5 | 0.08 |
| 2 | 200 | 63 | 6.1 | 0.16 |
| 3 | 300 | 95 | 4.5 | 0.22 |
| 4 | 400 | 125 | 2.0 | 0.50 |
| 5 | 500 | 154 | 2.0 | 0.50 |

The formulas for the estimates of $\alpha$ and $\beta$ remain the same:

$$\widehat{\beta} = \frac{s_{xy}}{s_{xx}} \qquad (6.2.12)$$

$$\widehat{\alpha} = \bar{y} - \widehat{\beta}\bar{x}. \qquad (6.2.13)$$

The estimate of $\sigma^2(x_i)$ is now

$$\widehat{\sigma}^2(x_i^2) = \frac{\widehat{K}_v}{w_i}, \qquad (6.2.14)$$

where $\widehat{K}_v$ is defined as

$$\widehat{K}_v = \frac{\sum_{i=1}^n w_i(y_i - \widehat{\alpha} - \widehat{\beta}x_i)^2}{n-2}. \qquad (6.2.15)$$

*Example 6.2.1: Calibration curve with variable $\sigma^2(x_i)$*
I decided to investigate how long it takes to boil water in my automatic coffee pot. The time depends on the amount of water. I poured 100, 200, 300, 400 and 500 cm$^3$ of water, and measured the boiling time in seconds. The results are presented in Table 6.2.

When I repeated the experiment, I noticed relatively large variations in the boiling time for a small amount of water, while for a large amount of water the boiling time remains almost constant. I decided to repeat the whole experiment in order to estimate the variance of the boiling time. The estimated variances $s_i^2$ are given in the fourth column of Table 6.2. The weights $w_i$ were taken as $w_i = 1/s_i^2$.

Figure 6.5 gives the Statistix printout for the weighted regression. The estimates are $\widehat{\alpha} = 4.4708$, $\widehat{\beta} = 0.29991$ and the estimate of $K_v$ is 0.22644.

```
WEIGHTED LEAST SQUARES LINEAR REGRESSION OF TIME

WEIGHTING VARIABLE: WEIGHT

PREDICTOR
VARIABLES     COEFFICIENT    STD ERROR    STUDENT'S T       P
---------     -----------    ---------    -----------    -----
CONSTANT        4.45037        1.33483          3.33      0.0446
WATER           0.29996        0.00335         89.46      0.0000


R-SQUARED                0.9996    RESID. MEAN SQUARE (MSE)    0.22942
ADJUSTED R-SQUARED       0.9995    STANDARD DEVIATION          0.47897


SOURCE         DF        SS            MS          F         P
----------     ---    ----------    ----------    -----    ------
REGRESSION       1    1836.02       1836.02     8003.03    0.0000
RESIDUAL         3    0.68825       0.22942
TOTAL            4    1836.71

CASES INCLUDED 5    MISSING CASES 0
```

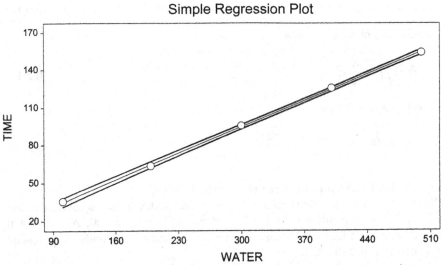

Figure 6.5. The printout and calibration curve for the data in Table 6.2 with 95% confidence belt.

## 6.2.2   Uncertainty in $x^\star$ for the Weighted Case

Now suppose that we observe a single value of the response $y^\star$. Then, exactly as in the "regular" (nonweighted) case, the estimate of $x^\star$ is obtained from the formula

$$x^\star = \overline{x} + \frac{y^\star - \overline{y}}{\widehat{\beta}}. \tag{6.2.16}$$

Note that in this formula, $\bar{x}, \bar{y}$ and $\hat{\beta}$ are computed from (6.2.7), (6.2.8) and (6.2.12).

After calculating $x^*$ from (6.2.16), we need to establish the uncertainty for $x^*$. Assume that we know the weight $w^*$ which corresponds to the calculated value of $x^*$. $w^*$ can be obtained by means of a specially designed experiment or, more simply, by interpolating between the weights $w_r$ and $w_{r+1}$ corresponding to the nearest neighbors of $x^*$ from the left and from the right.

The formula for the uncertainty in $x^*$ is similar to (6.1.21), with obvious changes following from introducing weights.

Note that the expression for $\mathrm{Var}[\hat{\beta}]$ is the same as (6.1.22), with obvious changes in the expression for $s_{xx}$; see, for example, Hald (1952, Chapt. 18, Sect. 6):

$$\mathrm{Var}[x^*] \approx (x^* - \bar{x})^2 \left[ \frac{1/w^* + 1/\sum_{i=1}^{n} w_i}{(y^* - \bar{y})^2} + \frac{1}{\hat{\beta}^2 s_{xx}} \right] \cdot \hat{K}_v. \tag{6.2.17}$$

Note that this formula reduces to (6.1.21) if we set all weights $w_i \equiv 1$.

*Example 6.2.1. continued.*
First compute $\bar{x} = 380.8$ and $\bar{y} = 118.7$ by (6.2.7) and (6.2.8). The value of $\hat{\beta}$ is in the printout: $\hat{\beta} = 0.29991$. Let $y^* = 75$ sec. Then it follows from (6.2.16) that $x^* = 235.2$. We take the corresponding weight $w^* = 0.18$, by interpolating between the weights 0.16 and 0.22. From the printout we know that $\hat{K}_v = 0.22644$. From Table 6.2 it follows that $\sum_{i=1}^{5} w_i = 1.46$ and $s_{xx} = 20,263$. Now

$$\mathrm{Var}[x^*] \approx (235.2 - 380.8)^2 \left[ \frac{1/0.18 + 1/1.46}{(75 - 118.7)^2} + \frac{1}{0.3^2 \cdot 20,263} \right] \cdot 0.2264$$
$$= 18.3.$$

Summing up, $x^* \pm \sqrt{\mathrm{Var}[x^*]} = 235.2 \pm 4.3$.

# 6.3 Calibration Curve When Both Variables Are Subject to Errors

## 6.3.1 Parameter Estimation

In this section we assume that $x$ is also subject to experimental error. This is a more realistic situation than the previously considered cases where the $x$s were assumed to be known without errors.

Consider, for example, the situation described in Example 6.1.1. In measuring absorbance, the solution of glucose is obtained by dissolving a certain amount of reagent in water. The exact amount of the reagent and the volume (or the weight) of the solution are known with error resulting from measurement errors in weighting and determining the volume. Thus the input value of $x$ fed into the spectrometer for calibration purposes is also subject to uncertainty, or to put it simply, to experimental error.

In this section our assumptions are the following:

1. The data is a set of pairs $[x_i, y_i, i = 1, 2, \ldots, n]$.

2. All values of $x_i$ are observations (sample values) of random variables $X_i$, with mean $E[X_i]$ and variance $\text{Var}[X_i] = \sigma_x^2$. Note that we treat here only the case of constant variances, i.e. $\sigma_x^2 = Const$.

3. All values of $y_i$ are observations (sample values) of random variables $Y_i$, with mean $E[Y_i]$ and variance $\text{Var}[Y_i] = \sigma_y^2$. Again, we consider only the case of constant variances $\sigma_y^2 = Const$.

4. $X_i, Y_i, i = 1, \ldots, n$, are mutually independent random variables.

5. The ratio of variances

$$\lambda = \frac{\sigma_y^2}{\sigma_x^2}. \tag{6.3.1}$$

is assumed to be a *known quantity*. A separate experiment must be carried out to estimate $\lambda$.

6. It is assumed that there is a *linear relationship* between the mean value of the input variable $X$ and the mean value of the output variable $Y$, i.e. we assume that

$$E[Y_i] = \alpha + \beta E[X_i]. \tag{6.3.2}$$

Our purpose is to find the "best" fit of our data to a linear relationship $y = \alpha + \beta x$. To proceed further, we need to put $X_i$ and $Y_i$ into similar conditions, i.e. balance the variances. This will be done by introducing a new random variable

$$Y_i^\star = \frac{Y_i}{\sqrt{\lambda}}. \tag{6.3.3}$$

Now obviously, $\text{Var}[Y_i^\star] = \text{Var}[X_i]$. Our "new" data set is now $[x_i, y_i^\star = y_i/\sqrt{\lambda}, i = 1, 2, \ldots, n]$.

It remains true that the mean values of $Y_i^\star$ and the mean value of $X_i$ are linearly related to each other:

$$E[Y_i^\star] = \alpha^\star + \beta^\star E[X_i]. \tag{6.3.4}$$

Obviously, $\alpha$ and $\beta$ are expressed through $\alpha^\star$ and $\beta^\star$ as

$$\alpha = \alpha^\star \sqrt{\lambda}, \quad \beta = \beta^\star \sqrt{\lambda}. \tag{6.3.5}$$

What is the geometry behind fitting a linear relationship to our data set? In Sect. 6.1, for the "standard" linear regression curve, we minimized the sum of squared distances of $y_i$ from the hypothetical regression line. There, $y$ and $x$ played different roles, since $x_i$ were assumed to be known without errors. Now the situation has changed, and both variables are subject to errors. The principle of fitting the data set to the linear calibration curve will now be *minimizing the sum of squares of the distances* of the points $(x_i, y_i^*)$ to the hypothetical calibration curve $y^* = \alpha^* + \beta^* x$. This is illustrated in Fig. 6.6.

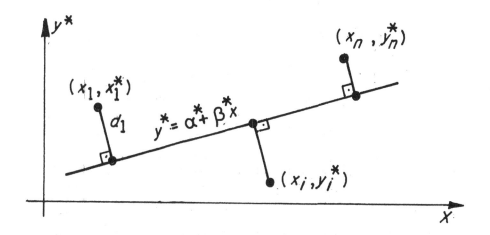

Figure 6.6. Constructing the calibration curve by minimizing $\sum d_i^2$

The formula expressing the sum of squares of the distances of the experimental points $(x_i, y_i^*)$ to the calibration curve $y^* = \alpha^* + \beta^* x$ is given by the following formula known from analytic geometry:

$$\psi(\alpha, \beta) = \sum_{i=1}^{n} \frac{(\alpha^* + \beta^* x_i - y_i^*)^2}{1 + (\beta^*)^2}.$$

Now substitute into this formula $y_i^* = y_i/\sqrt{\lambda}$. After simple algebra we obtain:

$$\psi(\alpha, \beta) = \sum_{i=1}^{n} \frac{(\sqrt{\lambda}\alpha^* + \sqrt{\lambda}\beta^* x_i - y_i)^2}{(1 + (\beta^*)^2)\lambda}. \tag{6.3.6}$$

The estimates of $\alpha^*$ and $\beta^*$, denoted as $\widehat{\alpha}^*$ and $\widehat{\beta}^*$ respectively, *minimize* the expression (6.3.6). To find them, we have to solve the set of two equations:

$$\partial \psi / \partial \alpha^* = 0, \quad \partial \psi / \partial \beta^* = 0. \tag{6.3.7}$$

Let us omit the technical details and present the final result. In the original coordinates $(x, y)$, the estimates of $\alpha$ and $\beta$ denoted as $\widehat{\alpha}$ and $\widehat{\beta}$ are as follows:

$$\widehat{\beta} = \frac{s_{yy} - \lambda s_{xx} + \sqrt{(\lambda s_{xx} - s_{yy})^2 + 4\lambda s_{xy}^2}}{2s_{xy}}, \tag{6.3.8}$$

$$\widehat{\alpha} = \bar{y} - \widehat{\beta}\bar{x}. \tag{6.3.9}$$

Table 6.3: Engine temperature in °C at two locations

| $i$ | $x_i = T_b(i)$ | $y_i = T_c(i)$ |
|-----|----------------|----------------|
| 1   | 73             | 79             |
| 2   | 132            | 83             |
| 3   | 211            | 121            |
| 4   | 254            | 159            |
| 5   | 305            | 167            |

Note that $\bar{x}, \bar{y}, s_{xx}, s_{xy}, s_{yy}$ are defined by (6.1.9)–(6.1.13), exactly as in the case of nonrandom $x$s.

The results (6.3.8) and (6.3.9) can be found in Mandel (1991, Chapt. 5). Standard texts on measurements and on regression, such as Miller and Miller (1993), typically do not consider the case of random errors in both variables.

*Remark 1.*
Suppose $\lambda \to \infty$. This corresponds to the situation when $x$s are measured without error. If we investigate the formula for $\widehat{\beta}$ when $\lambda$ goes to infinity, we obtain, after some algebra, the usual formula for $\widehat{\beta} = s_{xy}/s_{xx}$. The proof is left as an exercise.

*Example 6.3.1: Relationship between engine temperature at two locations*
An experiment is set up on engine testing bed to establish the relationship between the temperature $T_b$ of the crankshaft bearing and the temperature of the engine cap $T_c$. The usual engine temperature measurements are made only from the engine cap, and it is important to evaluate $T_b$ from the reading of $T_c$. Table 6.3 presents the measurement data obtained in the experiment.

It is assumed that there is a linear relationship between the true temperature at the bearing and the true temperature at the engine cap. It is assumed also that the variance of the temperature reading inside the engine is four times as large as the variance in measuring $T_c$. In our notation, it is assumed that $\lambda = 1/4 = 0.25$. Our goal is to estimate the parameters $\alpha$ and $\beta$ in the assumed linear relationship

$$E[T_c] = \alpha + \beta E[T_b]. \tag{6.3.10}$$

*Solution.* Let us return to familiar notation. Denote $x_i = T_b(i)$ and $y_i = T_c(i)$. Then it follows from Table 6.3 that

$$\bar{x} = 195.0, \bar{y} = 121.8, s_{xx} = 34\,690, s_{yy} = 6764.8, s_{xy} = 14\,820 \ .$$

Now by (6.3.8) and (6.3.9) we find that $\widehat{\alpha} = 36.0$ and $\widehat{\beta} = 0.44$. Thus, the best fit for the experimental data is the relationship

$$y = 36 + 0.44x, \text{ or } T_c = 36 + 0.44 \cdot T_b. \tag{6.3.11}$$

For example, the temperature on the cap is 180°C. Then solving (6.3.11) for $T_b$, we obtain $T_b = (180 - 36)/0.44 = 327.3°C$

## 6.3.2 Uncertainty in $x$ When Both Variables Are Subject to Errors

Suppose that we have a single observation on the response which we denote as $\tilde{y}$. Then the corresponding value of $x$, denoted as $\tilde{x}$, will be found from the calibration curve $y = \hat{\alpha} + \hat{\beta}x$ by the following formula:

$$\tilde{x} = \frac{\tilde{y} - \hat{\alpha}}{\hat{\beta}}. \tag{6.3.12}$$

Following Mandel (1991, Sect. 5.6), define the following quantities:

$$d_i = y_i - \hat{\alpha} - \hat{\beta}x_i; \tag{6.3.13}$$

$$D^2 = \sum_{i=1}^{n} d_i^2/(n-2); \tag{6.3.14}$$

$$P = \lambda + \hat{\beta}^2; \tag{6.3.15}$$

$$Q = \lambda^2 s_{xx} + 2\lambda\hat{\beta}s_{xy} + \hat{\beta}^2 s_{yy}. \tag{6.3.16}$$

We present the following result by Mandel (1991, p. 93):

$$\text{Var}[\tilde{x}] \approx \left[ \frac{D^2\lambda}{P\hat{\beta}^2} + \frac{D^2}{\hat{\beta}^2} \left( \frac{1}{n} + \frac{P^2(\tilde{y} - \bar{y})^2}{\hat{\beta}^2 Q} \right) \right]. \tag{6.3.17}$$

# 6.4 Exercises

**1.** Derive the formula (6.2.17).

**2.** The following table is based on data on carbon metrology given in Pankratz (1997).[1]

$x_i$ is the assigned carbon content in ppma (parts per million atomic) in specially prepared specimens. $y_i$ is the average carbon content computed from five measurements of the specimen $i$ carried out during the 6th day of the

---

[1] Source: Peter P. Pankratz "Calibration of an FTIR Spectrometer for Measuring Carbon", in the collection by Veronica Czitrom and Patrick D. Spagon *Statistical Case Studies for Industrial Process Improvement* ©1997. Borrowed with the kind permission of the ASA and SIAM.

experimentation (Pankratz 1997, p. 32). The sample variance $s_i^2$ was computed from these five measurements, and the weight was set up as $w_i = 0.001 \times (1/s_i^2)$.

| $i$ | $x_i$ | $y_i$ | Sample Variance | Weight $w_i$ |
|---|---|---|---|---|
| 1 | 0.0158 | 0.0131 | 0.000066 | 15 |
| 2 | 0.0299 | 0.0180 | 0.00008 | 12.5 |
| 3 | 0.371 | 0.3678 | 0.00022 | 4.5 |
| 4 | 1.1447 | 0.9750 | 0.00076 | 1.3 |
| 5 | 2.2403 | 2.1244 | 0.0027 | 0.4 |

Compute the parameters of the calibration line $y = \alpha + \beta x$ using weighted regression.

*Answer:* Using the weighted regression procedure in Statistix, we obtain the following estimates: $\hat{\alpha} = 0.0416$; $\hat{\beta} = 0.8827$; $\hat{K}_v = 0.15264$.

**3.** Suppose we observe $y^\star = 0.4$ in Exercise **2**. Find the corresponding value of $x^\star$ and the approximate value of $\sqrt{\text{Var}[x^\star]}$.

**4.** Investigate (6.3.8) when $\lambda \to \infty$ and prove that $\hat{\beta} \to ss_{xy}/s_{xx}$.

*Solution:* Represent the square root in (6.3.8) as $\lambda s_{xx}\sqrt{1-\gamma}$, where $\gamma = 2s_{yy}/(\lambda s_{xx}) - s_{yy}^2/(\lambda^2 s_{xx}^2) - 4s_{xy}^2/(\lambda s_{xx}^2)$, and use the approximation $\sqrt{1-\gamma} \approx 1 - \gamma/2$. Simplify the whole expression and let $\lambda \to \infty$.

**5.** Suppose that we observe, at each $x_i$, $k_i$ values of the response variable $y_i$ and use for fitting the regression line the data $[x_i, \overline{y_i}, i = 1, \ldots, n]$, where $\overline{y_i}$ is the average of $k_i$ observations. Argue that this case can be treated as a weighted regression with $w_i = k_i$.

**6.** Suppose that in Example 6.3.1 we observe $\tilde{y} = 180$. Compute $\text{Var}[\tilde{x}]$ using (6.3.17). Take $D^2 = 146.4$.

# Chapter 7

# Collaborative Studies

*The only way you can sometimes achieve a meeting of minds is by knocking a few heads together.*

14,000 Quips and Quotes for Writers and Speakers

## 7.1 Introduction: The Purpose of Collaborative Studies

So far we have dealt with various measurement problems which arise in a single measurement laboratory and which can be resolved within that laboratory. There often arise, however, measurement-related problems whose analysis and resolution involve *several* measurement laboratories. A typical example is the implementation of a new technology and/or product by enterprises which have different geographical locations. Suppose a pilot plant located in Texas, USA, develops a chemical process for producing new artificial fertilizer. An important property of this new product is that a certain chemical agent, which is responsible for underground water pollution, is present in very small amounts. Later on, the new process will be implemented in eight branches of the pilot plant which are located in different parts of the USA and elsewhere. The quality of the fertilizer produced in each branch will be controlled by the *local* laboratory.

We are faced with a new and unexpected phenomenon: all branches produce the fertilizer by the same technology, each branch has a seemingly stable production process, all laboratories follow the same instructions, but the measurement results obtained by different laboratories show a large disparity.

Miller and Miller (1993, p. 85) present two examples of such disparity: "In one study the level of polyunsaturated fatty acids in a sample of palm oil reported by 16 different laboratories varied from 5.5% to 15% .... Determination

Table 7.1: Results from collaborative tests reported by 11 labs

|   | 1 | 2 | 3 | 4 | 5 | 6 | 7 | 8 | 9 | 10 | 11 |
|---|---|---|---|---|---|---|---|---|---|----|----|
| $y$ | 16.0 | 16.1 | 16.3 | 17.1 | 16.5 | 16.5 | 16.7 | 16.9 | 16.5 | 16.5 | 16.5 |
| $x$ | 16.0 | 15.8 | 16.0 | 16.8 | 16.4 | 16.2 | 16.7 | 16.6 | 16.3 | 16.5 | 16.2 |

of the percentage of aluminium in a sample of limestone in ten laboratories produced values ranging from 1.11% to 1.9%."

In principle, there are two main reasons for the large disparity in the results: "usual" variations caused by measurement errors, and the presence of systematic errors. To clarify the second source of disparity, let us consider a fragment from Table 1 in Youden and Steiner (1975, p. 73).[1]

Table 7.1 presents measurement results of the same chemical characteristic made by 11 laboratories. Each received two specimens with the same amount of the chemical agent. Denote by $x$ and $y$ the smallest and the largest observation, respectively.

Youden and Steiner (1975) suggest to analyze the results by means of an $x - y$ scatter plot shown in Fig. 7.1. The $(x, y)$ points lie quite close to a line with slope 45°. The regression line of $y$ on $x$ is $y = 1.06 + 0.95x$.

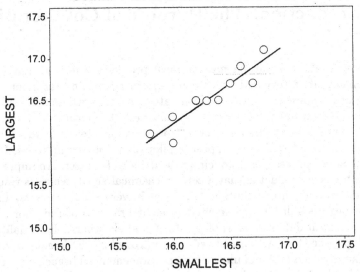

Figure 7.1. Scatter plot of $y$ versus $x$ for the data in Table 7.1.

---

[1] Reprinted, with permission, from the *Statistical Manual of the Association of Official Analytical Chemists*, (1975). Copyright, 1975, by AOAC INTERNATIONAL.

The explanation is that both measurement results of laboratory $i$ have the same *bias* $\delta_i$. Suppose that the true content is 16.2. Then most of the laboratories have a positive systematic error.

What is the reason for this systematic error? Probably, deviations by each local laboratory from the "standard" measurement procedure developed and recommended by the pilot plant laboratory.

The purpose of collaborative studies (CSs) is twofold. First, it is desirable to discover the factors which might cause the bias in the measurement process and eliminate them. This is mainly achieved by carrying out a specially designed experiment which will be described in the next section (the so-called "ruggedness" test). Second, analysis needs to be carried out to locate the laboratories which have excessive random and/or systematic errors. To achieve this goal, all laboratories taking part in the CS carry out measurements on a specially selected set of specimens. (All laboratories receive identical sets). An independent organization or the pilot laboratory performs the data analysis.

## 7.2 Ruggedness (Robustness) Test

After development of a new product and/or a new technology, the pilot laboratory which initiates the CS carries out a specially designed experiment in order to reveal the factors which in the course of the experiment might greatly influence the measurements of the new product parameters and cause large bias in the measurement results

Suppose that a technological process has been developed for producing a new chemical substance from plants. An important characteristic of this substance is its purity, mainly affected by small amounts of pesticides in it. Accurate measurement of the amount of pesticides is crucial for monitoring the production process.

Suppose that the pilot laboratory has established the following list of factors which may influence the measurement results: reagent purity (factor $A$); catalyst presence (factor $B$); reaction time (factor $C$); ambient temperature (factor $D$); air humidity (factor $E$); spectrograph type (factor $F$); and water quality (factor $G$).

The next step is crucial: a series of eight experiments are carried out with a special choice of factor combinations for each experiment; each factor appears at one of two levels shown in Table 7.2, denoted by either a capital or lower-case letter. These factor combinations are organized in a special way called *orthogonal design*, shown in Table 7.3. Thus, for example, experiment 4 is carried out using the following combination of factors: low reagent purity ($A$), no catalyst ($b$), reaction time 10 min ($c$), 20°C ambient temperature ($d$), etc. The observed result was $y_4 = 0.730$. Of course, in all experiments the measurements were carried out on a specimen with *the same* amount of impurity.

Table 7.2: Potential influencing factors on measurement results

| Factor | Description | Level 1 coding | Level 2 coding |
|--------|-------------|----------------|----------------|
| 1 | Reagent purity | Low $\Longrightarrow A$ | High $\Longrightarrow a$ |
| 2 | Catalyst presence | Yes $\Longrightarrow B$ | No $\Longrightarrow b$ |
| 3 | Reaction time | 20 min $\Longrightarrow C$ | 10 min $\Longrightarrow c$ |
| 4 | Ambient temp. | 25°C $\Longrightarrow D$ | 20°C $\Longrightarrow d$ |
| 5 | Air humidity | Low $\Longrightarrow E$ | High $\Longrightarrow e$ |
| 6 | Type of spectrograph | New type $\Longrightarrow F$ | Old type $\Longrightarrow f$ |
| 7 | Water quality | Regular $\Longrightarrow G$ | Distilled $\Longrightarrow g$ |

Table 7.3: Factor combination in orthogonal design

| Experiment $i$ | | | | | | | | Measurement result $y_i$ |
|----------------|---|---|---|---|---|---|---|---------------------------|
| 1 | $A$ | $B$ | $C$ | $D$ | $E$ | $F$ | $G$ | $y_1 = 0.836$ |
| 2 | $A$ | $B$ | $c$ | $D$ | $e$ | $f$ | $g$ | $y_2 = 0.858$ |
| 3 | $A$ | $b$ | $C$ | $d$ | $E$ | $f$ | $g$ | $y_3 = 0.776$ |
| 4 | $A$ | $b$ | $c$ | $d$ | $e$ | $F$ | $G$ | $y_4 = 0.730$ |
| 5 | $a$ | $B$ | $C$ | $d$ | $e$ | $F$ | $g$ | $y_5 = 0.770$ |
| 6 | $a$ | $B$ | $c$ | $d$ | $E$ | $f$ | $G$ | $y_6 = 0.742$ |
| 7 | $a$ | $b$ | $C$ | $D$ | $e$ | $f$ | $G$ | $y_7 = 0.823$ |
| 8 | $a$ | $b$ | $c$ | $D$ | $E$ | $F$ | $g$ | $y_8 = 0.883$ |

We assume that the following simple additive model of factor influence on the measurement result is valid:

$$
\begin{aligned}
Y_1 &= \mu + A + B + C + D + E + F + G + \epsilon_1, \\
Y_2 &= \mu + A + B + c + D + e + f + g + \epsilon_2, \\
Y_3 &= \mu + A + b + C + d + E + f + g + \epsilon_3, \\
Y_4 &= \mu + A + b + c + d + e + F + G + \epsilon_4, \\
Y_5 &= \mu + a + B + C + d + e + F + g + \epsilon_5, \\
Y_6 &= \mu + a + B + c + d + E + f + G + \epsilon_6, \\
Y_7 &= \mu + a + b + C + D + e + f + G + \epsilon_7, \\
Y_8 &= \mu + a + b + c + D + E + F + g + \epsilon_8.
\end{aligned}
\tag{7.2.1}
$$

Here $\mu$ is the overall mean value; the $\epsilon_i$ are assumed to be independent zero-mean measurement errors with variance $\sigma_0^2$. The special structure of equations (7.2.1) allows an extremely simple procedure for calculating the contribution of each of the factors. The rule is as follows. To obtain an estimate of the contribution of a single factor in the form of $A - a$, i.e. in the form of a difference in the measurement results when the factor $A$ is replaced by $a$, *add* all the $y_i$ containing the capital letter level of the factor A and *subtract* the $y_i$ which contain the small letter level of the factor. Then divide the result by 4. For example, factor $B$ is at Level 1 in experiments $1, 2, 5, 6$ and at Level 2 in experiments $3, 4, 7, 8$; see Table 7.3. Then

$$
\widehat{B - b} = \frac{y_1 + y_2 + y_5 + y_6 - (y_3 + y_4 + y_7 + y_8)}{4}.
\tag{7.2.2}
$$

Let us proceed with a numerical example.

*Example 7.2.1: Factors influencing the measurement results in Table 7.3*
The calculations reveal the following contributions of the factors:

$\widehat{A - a} = -0.0045$;

$\widehat{B - b} = -0.0015$;

$\widehat{C - c} = -0.002$;

$\widehat{D - d} = 0.096$;

$\widehat{E - e} = 0.014$;

$\widehat{F - f} = 0.005$;

$\widehat{G - g} = -0.039$.

The dot diagram in Fig. 7.2 shows that only two factors, $D$ and $G$ have a significant influence on the measurement result. The measurement result is especially sensitive to the ambient temperature. Increasing it by 5°C gives an increase in the result of $\approx 0.1$. So it is recommended to control carefully the

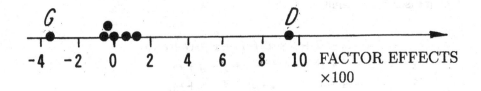

Figure 7.2. Dot diagram for the effects of factors $A, B, ..., G$.

ambient temperature and keep it near to 20°C. In addition, use of regular water
in the course of reaction causes a decrease in the result by 0.039. To reduce
disparity in measurement results from different laboratories, it is recommended
that all laboratories control this factor too and use only distilled water.

Another valuable conclusion from the above experiment is that the measurement results are robust (insensitive) with respect to the changes in the remaining
five factors.

The example considered here is the so-called $2^3$ factorial design. There is a
tremendous literature on using designed experiments to reveal the influencing
factors, including their possible interactions; see e.g. Montgomery (2001).

## 7.3    Elimination of Outlying Laboratories

Recommendations based on long experience say that the desirable number of
laboratories in the CS must be ten or more. Each laboratory must receive
three, four or five samples which correspond to different levels of the measurand
(say, concentration). In some situations, this level may be known exactly. All
laboratories receive identical sets of samples. It is advised for each laboratory
to carry out at least two independent measurements on each sample.

Table 7.4 summarizes the results of a CS carried out by 11 laboratories.
Each laboratory received three samples and measured each sample twice.[1]

We describe below a simple statistical procedure based on *ranking the sums
of replicates*. Its purpose is to identify the outlying laboratories, i.e. the laboratories with large systematic biases; see Youden and Steiner (1975)

**1.** Take the sum of both replicates for all laboratories and for all samples.
**2.** Rank the sums within one sample for all laboratories.
**3.** Compute the total rank of each laboratory as the sum of its ranks for each
sample.

---

[1]Reprinted from the *Statistical Manual of Official Analytical Chemists*, (1975). Copyright,
1975, by AOAC INTERNATIONAL

Table 7.4: Results from collaborative tests reported by 11 laboratories on three samples; Youden and Steiner 1975, p. 73

| Lab | Sample 1 | | Sample 2 | | Sample 3 | |
|-----|------|------|------|------|------|------|
| 1 | 21.1 | 21.4 | 12.7 | 12.9 | 16.0 | 16.0 |
| 2 | 21.4 | 21.6 | 13.2 | 13.0 | 16.1 | 15.8 |
| 3 | 20.8 | 20.7 | 13.1 | 12.8 | 16.3 | 16.0 |
| 4 | 21.9 | 21.6 | 13.5 | 13.1 | 17.1 | 16.8 |
| 5 | 21.0 | 20.9 | 12.9 | 13.0 | 16.5 | 16.4 |
| 6 | 20.9 | 20.4 | 12.8 | 12.7 | 16.5 | 16.2 |
| 7 | 21.2 | 20.9 | 12.8 | 12.7 | 16.7 | 16.7 |
| 8 | 22.0 | 21.1 | 13.0 | 12.9 | 16.6 | 16.9 |
| 9 | 20.7 | 21.0 | 12.6 | 12.9 | 16.3 | 16.5 |
| 10 | 20.9 | 21.3 | 12.1 | 12.8 | 16.5 | 16.7 |
| 11 | 21.1 | 20.6 | 13.0 | 12.8 | 16.5 | 16.2 |

Table 7.5: Ranking the sum of replicates from Table 7.4

| Lab | Sample 1 | rank | Sample 2 | rank | Sample 3 | rank | Total rank |
|-----|------|-----|------|-----|------|-----|-----|
| 1 | 42.5 | 8 | 25.6 | 5 | 32.0 | 2 | 15 |
| 2 | 43.0 | 9 | 26.2 | 10 | 31.9 | 1 | 20 |
| 3 | 41.5 | 2 | 25.9 | 8 | 32.3 | 3 | 13 |
| 4 | 43.5 | 11 | 26.6 | 11 | 33.9 | 11 | 33 |
| 5 | 41.9 | 5 | 25.9 | 8 | 32.9 | 7 | 20 |
| 6 | 41.3 | 1 | 25.5 | 3 | 32.7 | 4.5 | 8.5 |
| 7 | 42.1 | 6 | 25.5 | 3 | 33.4 | 9 | 18 |
| 8 | 43.1 | 10 | 25.9 | 8 | 33.5 | 10 | 28 |
| 9 | 41.7 | 3.5 | 25.5 | 3 | 32.8 | 6 | 12.5 |
| 10 | 42.2 | 7 | 24.9 | 1 | 33.2 | 8 | 16 |
| 11 | 41.7 | 3.5 | 25.8 | 6 | 32.7 | 4.5 | 14 |

Table 7.6: Criterion for rejecting a low or a high ranking laboratory score with 0.05 probability of wrong decision; Youden and Steiner 1975, p. 85

| No. of Laboratories | 3 Samples | 4 Samples | 5 samples |
|:---:|:---:|:---:|:---:|
| 6  | 3  18 | 5  23 | 7  28 |
| 7  | 3  21 | 5  27 | 8  32 |
| 8  | 3  24 | 6  30 | 9  36 |
| 9  | 3  27 | 6  34 | 9  41 |
| 10 | 4  29 | 7  37 | 10  45 |
| 11 | 4  32 | 7  41 | 11  49 |
| 12 | 4  35 | 7  45 | 11  54 |
| 13 | 4  38 | 8  48 | 12  58 |
| 14 | 4  41 | 8  52 | 12  63 |
| 15 | 4  44 | 8  56 | 13  67 |
| 16 | 4  47 | 9  59 | 13  71 |
| 17 | 5  49 | 9  63 | 14  76 |

Table 7.5 presents the ranking results of the data from Table 7.4. (In the case of tied ranks, each result obtains an average of all tied ranks.)

4. Compare the *highest and the lowest* rank with the lower and upper critical limits. These limits are given in Table 7.6. They are in fact 0.05 critical values, i.e. they might be exceeded by chance, in case of no consistent differences between laboratories, with probability 0.05.[2] If a certain laboratory shows a rank total lying outside one of the limits, this laboratory must be eliminated from the CS.

From the Table 7.5 we see that the total rank of laboratory 4 exceeds the critical value of 32 for 11 laboratories and three samples. Thus we conclude that this laboratory has a too large systematic error and we decide to exclude it from further studies. Note also that the $x - y$ plots (we presented only one in Figure 7.1, for Sample 3) point to the laboratory 4 as having the largest bias.

## 7.4   Homogeneity of Experimental Variation

Let us number the laboratories in the CS by $i = 1, \ldots, I$ and the samples by $j = 1, \ldots, J$. The *cell* $(i, j)$ contains $K$ observations, denoted as $x_{ijk}, k = 1, \ldots, K$. For the data in Table 7.4, we had at the beginning of the CS $I = 11$, $J = 3$, and $K = 2$ observations in each cell. After deleting laboratory 4, we preserve the numbers of the remaining laboratories but keep in mind that we now have 10 laboratories.

---

[2]Reprinted from the *Statistical Manual of Official Analytical Chemists*, (1975). Copyright, 1975, by AOAC INTERNATIONAL

Table 7.7: $k^*$-statistics for 10 laboratories and three samples

| Lab No | $R_{i,1}$ | $R_{i,2}$ | $R_{i,3}$ | $R_{i,1}/\overline{R}_1$ | $R_{i,2}/\overline{R}_2$ | $R_{i,3}/\overline{R}_3$ | Row sum |
|--------|-----------|-----------|-----------|--------------------------|--------------------------|--------------------------|---------|
| 1 | 0.3 | 0.2 | 0.0 | 0.83 | 0.87 | 0.0 | 1.70 |
| 2 | 0.2 | 0.2 | 0.3 | 0.56 | 0.87 | 1.5 | 2.93 |
| 3 | 0.1 | 0.3 | 0.3 | 0.28 | 1.30 | 1.5 | 3.08 |
| 5 | 0.1 | 0.1 | 0.1 | 0.28 | 0.43 | 0.5 | 1.21 |
| 6 | 0.5 | 0.1 | 0.3 | 1.39 | 0.43 | 1.5 | 3.33 |
| 7 | 0.3 | 0.1 | 0.0 | 0.83 | 0.43 | 0.0 | 1.27 |
| 8 | 0.9 | 0.1 | 0.3 | 2.50 | 0.43 | 1.5 | 4.43 |
| 9 | 0.3 | 0.3 | 0.2 | 0.83 | 1.30 | 1.0 | 3.14 |
| 10 | 0.4 | 0.7 | 0.2 | 1.11 | 3.04 | 1.0 | 5.15 |
| 11 | 0.5 | 0.2 | 0.3 | 1.39 | 0.87 | 1.5 | 3.76 |
| $\overline{R}_j$ | 0.36 | 0.23 | 0.2 | | | | |

Denote by $s_{ij}$ the standard deviation among replicates within cell $(i,j)$ and by $s_j$ the pooled value for replication standard deviation for sample $j$.

Mandel (1991, p. 176), suggested studying the experimental cell variability by using the so-called $k$-statistic which he defined as

$$k_{ij} = \frac{s_{ij}}{s_j}. \tag{7.4.1}$$

We will estimate $s_{ij}$ via the range of the replicates in the cell $(i,j)$, and $s_j$ via the average range of all cells in sample $j$. To compute $k_{ij}$, denote by $R_{i,j}$ the range of the replicates in the cell $(i,j)$, and by $\overline{R}_j$ the average range for sample $j$. For our case of $K = 2$ observations per cell, $R_{i,j}$ is the largest observation minus the smallest in the cell. The value of the $k$-statistic for cell $(i,j)$ we define as

$$k_{ij}^\star = \frac{R_{i,j}}{\overline{R}_j} = \frac{R_{i,j}}{\sum_{i=1}^{I} R_{i,j}/I}. \tag{7.4.2}$$

Columns , 5, 6, 7 of Table 7.7 present the values of $k_{ij}^\star$ for the data in Table 7.4.

For visual analysis of the data in Table 7.7 we suggest the display shown on Figure 7.3. The area (or the height) of the shaded triangle in the $(i,j)$th cell is proportional to the value of $k_{ij}^\star$. It is instructive to analyze the display column by column and row by row. The column analysis may give raise to the question whether the standard deviation in cell (8,1) or (10,2) is unusually high. This question may be answered on the basis of tables presented by Mandel (1991, pp. 237–239). One will find that the critical value of the $k_{ij}$-statistic for 10 laboratories and two replicates is 2.32, at the 1% significance level. Thus we may conclude that the within-cell variability in cells (8,1) and (10,2) is "too large". This does not mean that the measurement results should be ignored

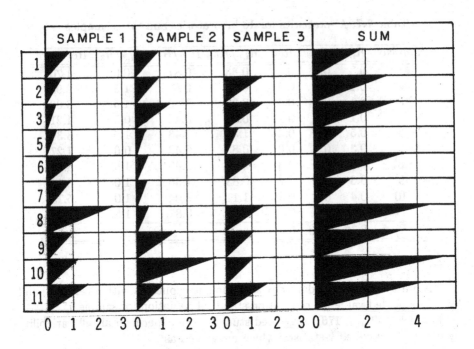

Figure 7.1. Display of $k_{ij}^\star$ statistics for data of Table 7.7.

and measurements repeated. These results must be marked for a more thorough investigation of the measurement conditions for the corresponding laboratories and samples.

The row-by-row analysis may serve as a comparison between laboratories. A row with high values of $k_{ij}^\star, j = 1, 2, 3$, demonstrates that laboratory $i$ has unusually high standard deviations. This is not a reason for excluding this laboratory from the CS, but rather a hint to investigate the potential sources of large within-cell variability in this laboratory. The last column in our display shows the row sums $\sum_{j=1}^{3} k_{ij}^\star$. For the data in the display, there is a suspicion that the measurement conditions in laboratories 8 and 10 cause the appearance of greater variability.

# Chapter 8

# Measurements in Special Circumstances

*The cause is hidden, but the result is known.*

Ovid, *Metamorphoses*

## 8.1 Measurements with Large Round-off Errors

### 8.1.1 Introduction and Examples

For sake of simplicity, assume that the measurement results are integers on a unit-step grid, i.e. the measurement results may be $\ldots, -2, -1, -2, -1, 0, 1, 2, \ldots$ In the notation of Sect. 2.2, the interval between instrument scale readings is $h = 1$.

Consider, for example, a situation where the measurand equals $\mu = 0.3$ and the measurement errors are normally distributed with mean zero and standard deviation $\sigma = 0.25$. So, the measurement result $Y$ (before the rounding off) is

$$Y = \mu + \epsilon, \tag{8.1.1}$$

where the measurement error $\epsilon \sim N(0, \sigma^2)$.

99.7% of the probability mass of $Y$ is concentrated in the interval $\mu \pm 3\sigma$, i.e. in the interval $[-0.45, 1.05]$. Taking into account the rounding-off, the probability of obtaining the result 0 is

$$\Phi((0.5 - \mu)/\sigma) - \Phi((-0.5 - \mu)/\sigma) = \Phi(0.8) - \Phi(-3.2)$$
$$= 0.7881 - 0.0007 = 0.7874, \tag{8.1.2}$$

and the probability of obtaining 1 is

$$\Phi((1.5 - \mu)/\sigma) - \Phi((0.5 - \mu)/\sigma) = \Phi(4.8) - \Phi(-0.8) \qquad (8.1.3)$$
$$= 1.0000 - 0.7881 = 0.2119.$$

It is easy to calculate that with probability 0.0007 the measurement result will be $-1$.

Let us denote by $Y^\star$ the measurement result *after* the rounding-off, i.e. the *observed* measurement result:

$$Y^\star = Y + \eta, \qquad (8.1.4)$$

where $\eta$ is the round-off error. In our example, all values of $Y$ which lie in the interval $(-0.5, 0.5]$ are automatically rounded to $Y^\star = 0$. All values of $Y$ which lie in the interval $(0.5, 1.5]$ are automatically rounded to $Y^\star = 1$ and all values of $Y$ in the interval $(-1.5, -0.5]$ produce $Y^\star = -1$.

In terms of Sect. 2.2, we have in this example a situation with *small $\sigma/h$* ratio: $\delta = \sigma/h = 0.25/1 = 0.25$. We termed this situation a *special* measurement scheme.

We consider in this section measurements with $\sigma/h \leq 0.5$, i.e. the situations with $\sigma \leq 0.5$. Our purpose remains estimating the measurand $\mu$ and the standard deviation $\sigma$ of the measurement error. The presence of large round-off errors (relative to the standard deviation of the measurement error $\sigma$) may invalidate our "standard" estimation methods. Let us consider a numerical example.

*Example 8.1.1: Bias in averaging*
Suppose, $\mu = 0.35$ and $\sigma = 0.25$. Then we obtain

$$p_0 = P(Y^\star = 0) = \Phi((0.5 - 0.35)/0.25) - \Phi((-0.5 - 0.35)/0.25)$$
$$= \Phi(0.6) - \Phi(-3.4) = 0.7257 - 0.0003 = 0.7254,$$
$$p_1 = P(Y^\star = 1) = \Phi((1.5 - 0.5)/0.25) - \Phi((0.5 - 0.35)/0.25)$$
$$= \Phi(4.0) - \Phi(0.6) = 1 - 0.7257 = 0.2743.$$

Thus the mean of the random variable $Y^\star$ is

$$E[Y] \approx 0.7254 \cdot 0 + 1 \cdot 0.2743 = 0.2743, \qquad (8.1.5)$$

which is *biased* with respect to the true value of $\mu = 0.35$.

Suppose that we carry out a series of $n$ independent measurements of the same measurand and $n$ is "large". Then $n_0 \approx n \cdot p_0$ occasions we will have the result $Y^\star = 0$ and on $n_1 \approx n \cdot p_1$ occasions we will have the result $Y^\star = 1$. Since $p_0 + p_1 \approx 1$, the average will be $\hat{\mu} = n_1/n \approx 0.2743$.

This is a biased estimate, and most striking is the fact that with the increase in the sample size $n$, the sample average approaches the mean of $Y^\star$, and thus the increase in the sample size *does not* reduce the bias. Formally speaking, our estimate of $\mu$ is not consistent. The standard remedy of increasing the sample size to obtain a more accurate estimate of the measurand simply does not work in our new situation with rounding-off.

The bias in this example is not very large. Suppose for a moment that $\mu = 0.4$ and $\sigma = 0.05$. It is easy to compute that practically all the probability mass of $Y^*$ is concentrated in two points, 0 and 1, and that $P(Y^* = 1) \approx 0.023$. This gives the mean of $Y^* \approx 0.023$. The estimate of $\mu$ is biased by 0.38.

Consider another example related to estimating $\sigma$.

*Example 8.1.2: Bias in estimating $\sigma$*
Consider $Y \sim N(\mu = 0.5, \sigma = 0.25)$. Practically all measurement results of $Y$ are concentrated on the support $[\mu - 3 \cdot \sigma, \mu + 3 \cdot \sigma] = [-0.25, 1.25]$, and we observe $Y^*$ equal to either 0 or 1 with probabilities 0.5.

Consider using ranges to estimate the standard deviation. The theory which is the basis for obtaining the coefficients $A_n$ in Table 2.6 presumes that the data are taken from a normal sample.

Suppose we have a sample of $n = 5$. The probability that all measurement results will be either $Y^* = 0$ or $Y^* = 1$ is $2 \cdot (0.5)^5 = 0.0625$. This is the probability of having the estimate of $\sigma$ equal zero. With complementary probability 0.9375 the observed range will be 1, and in that case the estimate $\hat{\sigma} = 1/A_5 \approx 0.43$. On average, our estimate will be $0 \cdot 0.0625 + 0.43 \cdot 0.9375 \approx 0.4$, which is considerably biased with respect to the true value 0.25.

We see, therefore, that rounding-off invalidates the range method, which theoretically, in the absence of round-off, produces unbiased estimates.

Let us try another estimate of $\sigma$ based on computing the variance of the two-point discrete distribution of the observed random variable $Y^*$.

Suppose we have $n = 5$ replicates of $Y^*$. With probability 2/32 we will observe either 0 on five occasions or 1 on five occasions. The corresponding variance estimate is zero. With probability 10/32, we will observe $(4, 1)$ or $(1, 4)$, and the variance estimate will be 4/25. With probability 20/32, we will observe $(3, 2)$ or $(2, 3)$ and the variance estimate will be 6/25. Thus, the expected value of the estimate will be

$$0 \cdot (2/32) + (4/25) \cdot (10/32) + (20/32) \cdot (6/25) = 0.2.$$

The estimate of the standard deviation is $\hat{\sigma} = \sqrt{0.2} = 0.447$, almost twice the true standard deviation.

## 8.1.2   Estimation of $\mu$ and $\sigma$ by Maximum Likelihood Method: $Y^* = 0$ or 1

Let the total number of observations be $n$, $n = n_0 + n_1$, where $n_0$ and $n_1$ are the numbers of observations of the points 0 and 1, respectively. Suppose that all measurements produce the same result, say $n = n_0$. Then our only conclusion will be that the estimate of the mean is $\hat{\mu} = 0$.

Regarding $\sigma$, we can say only that the length of the support of the random variable $\epsilon$ does not exceed $h = 1$. In practical terms, this means that $\sigma < 1/\sqrt{12}$.

From the measurement point of view, the instrument in that case is not accurate enough and should be replaced by one which has a finer scale (i.e. smaller value of $h$).

Now let us consider the case where the observations are located at two points, i.e. both $n_0 \geq 1$ and $n_1 \geq 1$. Our entire information is, therefore, two numbers $n_0$ and $n_1$.

### Estimation of $\mu$ When $\sigma$ Is Known

Assume $\sigma$ is known from previous measurement experience. Note that there is always a "naive" estimator of $\mu$ based on relative frequencies. We denote it $\mu_n$:

$$\mu_n = (0 \cdot n_0 + 1 \cdot n_1)/n = n_1/n. \tag{8.1.6}$$

We believe that the following *maximum likelihood* estimate of $\mu$, denoted as $\mu_{ML}$, has better properties than $\mu_n$.

**Step 1.** Write down the so-called *likelihood function* which in fact is the probability of observing the sample which we already have:

$$Lik = [P(Y^\star = 0)]^{n_0} \cdot [P(Y^\star = 1)]^{n_1}, \tag{8.1.7}$$

where

$$P(Y^\star = 0) = \Phi\big((0.5 - \mu)/\sigma\big) - \Phi\big((-0.5 - \mu)/\sigma\big), \tag{8.1.8}$$
$$P(Y^\star = 1) = \Phi\big((1.5 - \mu)/\sigma\big) - \Phi\big((0.5 - \mu)/\sigma\big).$$

Note that $Lik$ depends on $\mu$ only since $\sigma$ is assumed to be known.

**Step 2.** Find the maximum likelihood estimate of $\mu$. The maximum likelihood estimate $\mu_{ML}$ is that value of $\mu$ which *maximizes* the likelihood function, see DeGroot (1975, p. 338). In practice, it is more convenient to maximize the logarithm of the likelihood function (which will produce the same result):

$$\mu_{ML} = Arg \max_{\mu}[Log[Lik]]. \tag{8.1.9}$$

The estimation of $\mu_{ML}$ needs special computer software. Below we demonstrate an example by using Mathematica; see Wolfram (1999).

*Example 8.1.3: Maximum likelihood estimate of $\mu$*

Suppose that our data are $n_0 = 4$, $n_1 = 1$. From previous experience it is assumed that $\sigma = 0.3$. All computations are presented in the printout in Fig. 8.1. We see, therefore, that $\mu_{ML} = 0.256$. Note that the naive estimate is $\mu_n = 0.2$.

```
In[1]:= << Statistics`ContinuousDistributions`
        ndist = NormalDistribution[0, 1];
        σ = 0.3;
        n0 = 4; n1 = 1;
        p0 = CDF[ndist, (0.5 - μ) / σ] - CDF[ndist, (-0.5 - μ) / σ];
        p1 = CDF[ndist, (1.5 - μ) / σ] - CDF[ndist, (0.5 - μ) / σ];
        Lik = (p0)^n0 * (p1)^n1;
        FindMinimum[-Log[Lik], {μ, 0.4}]
        Plot[Log[Lik], {μ, 0.1, 0.7}]

Out[7]= {2.53282, {μ → 0.256461}}
```

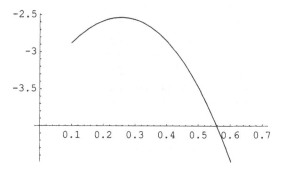

Figure 8.1. Mathematica printout of program for computing $\mu_{ML}$

### Maximum Likelihood When Both $\mu$ and $\sigma$ Are Unknown

Our data remain the same as in the previous case: we observe the result $Y^\star = 0$ $n_0$ times, and $Y^\star = 1$  $n_1 = n - n_0$ times, $n_0 > 0, n_1 > 0$.

Let us first try the method of maximum likelihood also in the case of two unknown parameters. The maximum likelihood function has the same form as (8.1.7), only now both $\mu$ and $\sigma$ are unknown:

$$Lik = [\Phi((0.5 - \mu)/\sigma) - \Phi((-0.5 - \mu)/\sigma)]^{n_0} \qquad (8.1.10)$$
$$\times[\Phi((1.5 - \mu)/\sigma) - \Phi((-0.5 - \mu)/\sigma)]^{n_1}.$$

Now we have to find the maximum of $Log[Lik]$ with respect to $\mu$ and $\sigma$.

The numerical investigation of the likelihood function (8.1.10) has revealed that the surface $Lik = \psi(\mu, \sigma)$ has a long and flat ridge in the area of the maximum. This can be well seen from the contour plot of $\psi(\mu, \sigma)$ in Fig. 8.2. In that situation, the "FindMinimum" operator produces results which, in fact, are not minimum points. Moreover, these results depend on the starting point of the minimum search.

```
<< Statistics`ContinuousDistributions`
ndist = NormalDistribution[0, 1];
n0 = 3; n1 = 7;
p0 = CDF[ndist, (0.5 - μ) / σ] - CDF[ndist, (-0.5 - μ) / σ];
p1 = CDF[ndist, (1.5 - μ) / σ] - CDF[ndist, (0.5 - μ) / σ];
Lik = (p0)^n0 * (p1)^n1;
ContourPlot[Log[Lik], {μ, 0.48, 0.72}, {σ, 0.05, 0.25},
    Contours → 39]
```

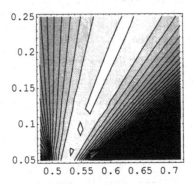

Figure 8.2. The contour plot of the likelihood function (8.1.10); $n_0 = 3, n_1 = 7$

For example, when the starting point is $(\mu = 0.6, \sigma = 0.1)$, Mathematica finds $(\mu_{ML} = 0.55, \sigma_{ML} = 0.1)$; when the starting point is $(\mu = 0.7, \sigma = 0.2)$, the results are $(\mu_{ML} = 0.59, \sigma_{ML} = 0.18)$. The values of $\psi(\mu_{ML}, \sigma_{ML})$ coincide in these two cases up to six significant digits.

What we suggest to do in this unfortunate situation is to use for $\sigma$ a naive frequency estimator with the Sheppard correction; see Sect. 2.2:

$$\tilde{\sigma}_n = \sqrt{\frac{n_0 \cdot n_1}{n \cdot n} - \frac{1}{12}}. \qquad (8.1.11)$$

Afterwards, substitute $\tilde{\sigma}_n$ into the expression (8.1.10) and optimize the likelihood function with respect to $\mu$ only. This becomes a well-defined problem. We present in Table 8.1 the calculation results in a ready to use form for all cases of $n = 10$ measurements with $n_0 > 0, n_1 > 0$.

## 8.1.3 Estimation of $\mu$ and $\sigma$: $Y^\star = 0, 1, 2$

Suppose that the measurand $\mu = 0.95$, and the measurement error $\epsilon \sim N(0, \sigma^2)$, with $\sigma = 0.45$. Then the $\pm 3\sigma$ zone around $\mu$ is $[-0.4, 2.3]$. Then three values of $Y^\star$ are likely to be observed: $0, 1$ or $2$. Suppose that we made $n$ measurements, observing $Y^\star = 0$ on $n_0$ occasions, $Y^\star = 1$ on $n_1$ occasions and $Y^\star = 2$ on

Table 8.1: Estimates $\mu_{ML}$ and $\tilde{\sigma}_n$ for $n_0 + n_1 = 10$

| $n_0$ | $n_1$ | $\mu_{ML}$ | $\tilde{\sigma}_n$ |
|---|---|---|---|
| 1 | 9 | 0.605 | 0.082 |
| 2 | 8 | 0.731 | 0.28 |
| 3 | 7 | 0.675 | 0.36 |
| 4 | 6 | 0.591 | 0.40 |
| 5 | 5 | 0.50 | 0.41 |
| 6 | 4 | 0.409 | 0.40 |
| 7 | 3 | 0.325 | 0.36 |
| 8 | 2 | 0.269 | 0.28 |
| 9 | 1 | 0.395 | 0.082 |

$n_2 = n - n_0 - n_1$ occasions, such that $n_0, n_1, n_2$ are all nonzero. Our purpose remains to estimate $\mu$ and $\sigma$.

Let us present first the naive estimates based on the first two moments of the discrete random variable $Y^\star$:

$$E[Y^\star] = 1 \cdot P(Y^\star = 1) + 2 \cdot P(Y^\star = 2). \tag{8.1.12}$$

Replacing the probabilities by the relative frequencies, we arrive at the following estimate of $\mu$:

$$\tilde{\mu} = n_1/n + 2n_2/n. \tag{8.1.13}$$

For the estimate of $\sigma$ we suggest the estimate of the standard deviation of $Y^\star$ with the Sheppard correction; see Sect. 2.2. The variance of $Y^\star$ equals

$$\text{Var}[Y^\star] = E[(Y^\star)^2] - (E[Y^\star])^2. \tag{8.1.14}$$

¿From here we arrive at the following estimate of $\sigma$:

$$\tilde{\sigma} = \sqrt{1^2 \cdot n_1/n + 2^2 \cdot n_2/n - (n_1/n + 2 \cdot n_2/n)^2 - 1/12}. \tag{8.1.15}$$

The maximum likelihood estimates of $\mu$ and $\sigma$ are obtained by maximizing with respect to $\mu$ and $\sigma$ the following expression for the likelihood function:

$$Lik = [P(Y^\star = 0)]^{n_0} \cdot [P(Y^\star = 1)]^{n_1} \cdot [P(Y^\star = 2)]^{n_2}, \tag{8.1.16}$$

where

$$P(Y^\star = 0) = \Phi\left(\frac{0.5 - \mu}{\sigma}\right) - \Phi\left(\frac{-0.5 - \mu}{\sigma}\right), \tag{8.1.17}$$

$$P(Y^\star = 1) = \Phi\left(\frac{1.5 - \mu}{\sigma}\right) - \Phi\left(\frac{0.5 - \mu}{\sigma}\right), \tag{8.1.18}$$

$$P(Y^\star = 0) = \Phi\left(\frac{2.5 - \mu}{\sigma}\right) - \Phi\left(\frac{1.5 - \mu}{\sigma}\right). \tag{8.1.19}$$

Table 8.2: Estimates of $\mu$ and $\sigma$ for $n_0 + n_1 + n_2 = 10$

| $n_0$ | $n_1$ | $n_2$ | $\mu_{ML}$ | $\sigma_{ML}$ | $\tilde{\mu}$ | $\tilde{\sigma}$ |
|---|---|---|---|---|---|---|
| 3 | 6 | 1 | 0.800 | 0.52 | 0.800 | 0.53 |
| 2 | 7 | 1 | 0.898 | 0.46 | 0.900 | 0.45 |
| 3 | 5 | 2 | 0.901 | 0.63 | 0.900 | 0.64 |
| 2 | 6 | 2 | 1.000 | 0.56 | 1.000 | 0.56 |

An interesting empirical fact is that the maximum likelihood estimates of $\mu_{ML}$ and $\sigma_{ML}$ practically coincide with the naive estimates $\tilde{\mu}, \tilde{\sigma}$, as Table 8.2 shows.

## 8.2  Measurements with Constraints

We will consider this subject on the basis of an example involving measuring angles in a triangle.

Suppose that we made independent measurements of the angles $\alpha, \beta$ and $\gamma$ in a triangle. We have obtained $\alpha_1, \alpha_2, \ldots, \alpha_k$ for angle $\alpha$; $\beta_1, \beta_2, \ldots, \beta_k$ for angle $\beta$, and $\gamma_1, \gamma_2, \ldots, \gamma_k$ for angle $\gamma$.

Of course our estimates of the angles $\widehat{\alpha}, \widehat{\beta}, \widehat{\gamma}$ must satisfy the natural constraint

$$\widehat{\alpha} + \widehat{\beta} + \widehat{\gamma} = 180°. \tag{8.2.1}$$

The approach to finding the "best" estimates of the angles is the following: we are looking for such values of $\widehat{\alpha}, \widehat{\beta}$ and $\widehat{\gamma}$ which *minimize* the expression

$$\sum_{i=1}^{k} \left(\alpha_i - \widehat{\alpha}\right)^2 + \sum_{i=1}^{k} \left(\beta_i - \widehat{\beta}\right)^2 + \sum_{i=1}^{k} \left(\gamma_i - \widehat{\gamma}\right)^2 \tag{8.2.2}$$

*subject* to the constraint (8.2.1).

In order to take the constraint into account let us express $\widehat{\gamma}$ from (8.2.1) as $\widehat{\gamma} = 180 - \widehat{\alpha} - \widehat{\beta}$ and substitute it into (8.2.2). Then our task is to minimize the expression

$$D(\widehat{\alpha}, \widehat{\beta}) = \sum_{i=1}^{k} \left(\alpha_i - \widehat{\alpha}\right)^2 + \sum_{i=1}^{k} \left(\beta_i - \widehat{\beta}\right)^2 + \sum_{i=1}^{k} \left(\gamma_i - (180 - \widehat{\alpha} - \widehat{\beta})\right)^2. \tag{8.2.3}$$

To solve this problem, we have to solve the following system of equations:

$$\partial D / \partial \widehat{\alpha} = 0, \ \partial D / \partial \widehat{\beta} = 0. \tag{8.2.4}$$

After simple algebra, these equation take the following form:

$$2\widehat{\alpha} + \widehat{\beta} = 180 - \overline{\gamma} + \overline{\alpha}, \tag{8.2.5}$$
$$\widehat{\alpha} + 2\widehat{\beta} = 180 + \overline{\beta} - \overline{\gamma}, \tag{8.2.6}$$

where $\overline{\alpha} = \sum_{i=1}^{k} \alpha_i/k$, $\overline{\beta} = \sum_{i=1}^{k} \beta_i/k$ and $\overline{\gamma} = \sum_{i=1}^{k} \gamma_i/k$ are the respective average values.

It is easy to solve these equations and to obtain that

$$\widehat{\alpha} = 60 - \overline{\gamma}/3 - \overline{\beta}/3 + 2\overline{\alpha}/3, \tag{8.2.7}$$

$$\widehat{\beta} = 60 - \overline{\gamma}/3 + 2\overline{\beta}/3 - \overline{\alpha}/3. \tag{8.2.8}$$

These two equations must be complemented by a third one which is obtained by substituting the estimates for $\alpha$ and $\beta$ into the constraint. This equation is

$$\widehat{\gamma} = 60 + 2\overline{\gamma}/3 - \overline{\beta}/3 - \overline{\alpha}/3. \tag{8.2.9}$$

Let us now estimate the variance of the angle estimates. Assume that all angles are measured with errors having equal variances $\sigma^2$. Then

$$Var[\overline{\alpha}] = \sigma^2/k = Var[\overline{\beta}] = Var[\overline{\gamma}]. \tag{8.2.10}$$

Since the measurements are independent, it follows from from (8.2.7) that

$$Var[\widehat{\alpha}] = Var[\overline{\gamma}]/9 + Var[\overline{\beta}]/9 + 4Var[\overline{\alpha}]/9. \tag{8.2.11}$$

Using (8.2.10), we obtain that

$$Var[\widehat{\alpha}] = \frac{2\sigma^2}{3k}. \tag{8.2.12}$$

By symmetry, the variances of $\widehat{\beta}$ and $\widehat{\gamma}$ are the same.

It is worth noting that the individual measurements $\alpha_i, \beta_i, \gamma_i$ do not appear in the results, only the averages $\overline{\alpha}, \overline{\beta}$ and $\overline{\gamma}$. We would obtain exactly the same result if the expression (8.2.3) were written as

$$\left(\overline{\alpha} - \widehat{\alpha}\right)^2 + \left(\overline{\beta} - \widehat{\beta}\right)^2 + \left(\overline{\gamma} - \widehat{\gamma}\right)^2, \tag{8.2.13}$$

and minimized subject to the same constraint (8.2.1). This fact has the following geometric interpretation. We minimize the distance from the point $(\overline{\alpha}, \overline{\beta}, \overline{\gamma})$ to the hyperplane $\widehat{\alpha} + \widehat{\beta} + \widehat{\gamma} = 180°$.

## 8.3  Exercises

1. In measuring the triangle angles, the following results were obtained:

$\alpha_1 = 30°30'; \alpha_2 = 29°56';$
$\beta_1 = 56°25'; \beta_2 = 56°45';$
$\gamma_1 = 93°50'; \gamma_2 = 94°14'.$

Find the "best" estimates of $\alpha, \beta, \gamma$.

2. For the data of 1, find the estimate of $\sqrt{\mathrm{Var}[\hat{\alpha}]}$.

3. In measuring the ohmic resistance of a specimen using a digital instrument, the following results (in ohms) were obtained:

10.1, 10.2, 10.2, 10.1, 10.1.

It is known that the instrument has measurement error $\epsilon \sim N(0, \sigma^2)$ with $\sigma = 0.03$. Find the maximum likelihood estimate $\mu_{ML}$ of the ohmic resistance.

*Hint.* Multiply all measurement results by a factor of 10. This will also increase $\sigma$ by a factor of 10. Put the zero of the scale at the point 101. Then there are three measurements equal to zero, $n_0 = 3$, and two equal to 1, $n_1 = 2$. Now we are in the situation of Example 8.1.2. On the new scale $\mu_{ML} = 0.256$, and in the original scale the estimate of the resistance is $10.1 + 0.0256 \approx 10.13$.

4. Suppose that the random variable $X \sim U(-0.7, 1.3)$. The measurements are made on a grid $\ldots, -2, -1, 0, 1, 2, \ldots$ with corresponding round-off. Find the mean and the standard deviation of the observed measurement results.

5. Five weight measurements are made on a digital instrument with scale step $h = 1$ gram, and the following results obtained: 110 gram, 111 gram (3 times) and 112 gram. Suppose that the weight is an uknown constant $\mu$ and the measurement error has a normal distribution $N(0, \sigma^2)$. Find the maximum likelihood estimates of $\mu$ and $\sigma$.

# Answers and Solutions to Exercises

**Chapter 2.**

**1, a.** $\hat{\mu} = 0.634$;

**1, b.** $s = 0.017$; the estimate of $\sigma$ using the sample range is $\hat{\sigma} = 0.06/3.472 = 0.017$;

**1, c.** $Q = (0.61 - 0.60)/0.06 = 0.167 < 0.338$. The minimal sample value is not an outlier.

**1, d.** The 0.95 confidence interval on $\mu$ is $[0.6246, 0.6434]$

**1, e.** $h = 0.01$, $s/h = 1.7$. It is satisfactory

**2.** *Solution.* $Y \sim N(150, \sigma = 15)$. $P(145 < Y < 148) = P((145 - 150)/15 < (Y - 150)/15 < (148 - 150)/15) = \Phi(-0.133) - \Phi(-0.333) = \Phi(0.333) - \Phi(0.133) = 0.630 - 0.553 = 0.077$.

**Chapter 3**

**1.** The sample averages are $\bar{x}_1 = 3200.1$, $\bar{x}_4 = 3359.6$. The sample variances are $s_1^2 = 2019.4$, $s_4^2 = 5984.6$. Assuming equal variances, the $T$-statistic for testing $\mu_1 = \mu_4$ against $\mu_1 < \mu_4$ equals $-3.99$, which is smaller than $t_{0.005}(8)$. We reject the null hypothesis.

**2.** For Table 4.1 we obtain the maximal variance for day 4, $s_4^2 = 52.29$, and the minimal $s_1^2 = 8.51$ for day 1. Hartley's statistic equals 6.10, which is less than the 0.05 critical value of 33.6 for 4 samples and 7 observations in each sample. The null hypothesis is not rejected.

**3.** For Table 3.8 data, the 95 % confidence interval on mean difference is $[0.02, 2.98]$. Since it does not contain zero, we reject the null-hypothesis.

**Chapter 4**

**1.** The average range per cell is 0.0118. It gives $\hat{\sigma}_e^* = 0.0070$. This is in close agreement with $\hat{\sigma}_e = 0.0071$ obtained in Sect. 4.4.

**3.** $\hat{\sigma}_X = 0.092$, $\hat{\sigma}_{e1} = 2.8 \cdot 10^{-2}$; $\hat{\sigma}_{e2} = 2.1 \cdot 10^{-2}$. The estimate of $\delta$ is $\hat{\delta} = -0.21$.

**4.** The 95 % confidence interval on $\delta$ is $[0.194, 0.227]$.

**7.** $F_A = (791.6/4)/(234.0/5) = 4.23$, which is less than $\mathcal{F}_{4,5}(0.05) = 5.19$. Thus we do not reject the null hypothesis at the significance level 0.05 but we do reject it at the level 0.1 for which $\mathcal{F}_{4,5}(0.05) = 3.52$.

$F_B = (234/5)/(0.98/10) = 477.6 \gg \mathcal{F}_{5,10}(0.05) = 3.33$. Thus the null hypothesis is certainly rejected.

## Chapter 6

**1.** Exactly as in the case of constant variance,

$$\mathrm{Var}[x^\star] \approx (x^\star - \bar{x})^2 \left[ \frac{\mathrm{Var}[y^\star] + \mathrm{Var}[\bar{y}]}{(\bar{y} - y^\star)^2} + \frac{\mathrm{Var}[\hat{\beta}]}{\hat{\beta}^2} \right].$$

Now use the following formulas for variance estimates:

$$\widehat{\mathrm{Var}}[y^\star] = \hat{K}_v / w^\star;$$
$$\widehat{\mathrm{Var}}[\bar{y}] = |\mathrm{by}\ (6.2.8)| = \left( \sum_{i=1}^n w_i^2 \hat{K}_v / w_i \right) / \left( \sum_{i=1}^k w_i \right)^2 = \hat{K}_v / \sum_{i=1}^n w_i;$$
$$\widehat{\mathrm{Var}}[\hat{\beta}] = \hat{K}_v / s_{xx}.$$

and substitute them in the expression for $\mathrm{Var}[x^\star]$.

**3.** *Solution.* $\bar{x} = 0.0938$; $\bar{y} = 0.1244$; $s_{xx} = 3.4355$; $s_{xy} = 3.0324$; $s_{yy} = 3.1346$, $\sum_{i=1}^5 w_i = 33.7$. $x^\star = (0.4 - 0.0416)/0.8827 = 0.406$.

To obtain the weight $w^\star$ for $x^\star$ interpolate linearly between $w_3$ and $w_4$. $w^\star \approx 4.4$. Then using (6.2.17) with $\hat{K}_v = 0.15264$, obtain that

$$\mathrm{Var}[x]^\star \approx (0.406 - 0.0938)^2 \left( \frac{1/4.4 + 1/33.7}{(0.4 - 0.1244)^2} + 1/(0.8827^2 \cdot 3.4355) \right) \cdot 0.15264 =$$
0.0559. Thus the result is: $0.406 \pm 0.236$.

**5.** If $\bar{y}_i = (y_{i1} + \ldots + y_{ik_i})/k_i$, then $\mathrm{Var}[\bar{y}_i] = \mathrm{Var}[y]_{i1}/k_i$. Compare this with (6.2.3): $w_i = k_i$.

**6.** Compute from the data in Table 6.3 that $D^2 = \sum_{i=1}^5 (y_i - 36 - 0.44x_i)^2/3 = 146.4$. Use (6.3.15) and (6.3.16) to calculate that $P = 0.444$ and $Q = 6738.2$. Substitute the values of $D^2, \lambda, P, \hat{\beta}, Q, \hat{y} = 180$ and $\bar{y}$ into (6.3.17). The result is $\mathrm{Var}[\tilde{x}] \approx 964.1$.

## Chapter 8

**1.** *Answer:* $\hat{\alpha} = 29°56'$; $\hat{\beta} = 56°18'$.

**2.** *Solution.* $\widehat{\mathrm{Var}}[\alpha] = (\alpha_1 - \alpha_2)^2/2 = 34^2/2 = 578$. Similarly, $\widehat{\mathrm{Var}}[\beta] = 20^2/2 = 200$, and $\widehat{\mathrm{Var}}[\gamma] = 24^2/2 = 288$.

The pooled variance estimate of $\sigma^2$ is $\hat{\sigma}^2 = (578 + 200 + 288)/3 = 355.3$. By (8.2.12), $\text{Var}[\hat{\alpha}] = \sigma^2/3$ and $\widehat{\text{Var}}[\hat{\alpha}] = 355.3/3 = 118.4$. $\sqrt{\widehat{\text{Var}}[\alpha]} = 11'$.

4. *Solution.* With probability 0.1 the result will be $-1$, with probability 0.5 it will be zero, and with the remaining probability 0.4 it will be $+1$. So, the measurement result will be a discrete random variable $Y^\star$, with $E[Y^\star] = 0.1(-1) + 0.5 \cdot 0 + 1 \cdot 0.4 = 0.3$. $Var[Y^\star] = 0.1 \cdot (-1)^2 + 0.4 \cdot (1^2) - 0.3^2 = 0.5 - 0.09 = 0.41$.

5. *Solution.* Consider expression (8.1.15) with $n_0 = 2, n_1 = 6, n_2 = 2$. Set the zero at 110 gram. This case is presented in Table 8.2, and the results are $\mu_{ML} = 1, \sigma_{ML} = 0.56$. Note that the likelihood function for the data $n_0 = 1, n_1 = 3, n_2 = 1$ is the square root of the likelihood function (8.1.15), and thus it will be maximized at the same values. Finally, $\mu_{ML} = 111, \sigma_{ML} = 0.56$ gram.

# Appendix A: Normal Distribution

$$\Phi(x) = \frac{1}{\sqrt{2\pi}} \int_{-\infty}^{x} \exp[-v^2/2]dv$$

Hundredth parts of $x$

| $x$ | 0 | 1 | 2 | 3 | 4 | 5 | 6 | 7 | 8 | 9 |
|-----|-----|-----|-----|-----|-----|-----|-----|-----|-----|-----|
| 0.0 | 0.5000 | 0.5040 | 0.5080 | 0.5120 | 0.5160 | 0.5199 | 0.5239 | 0.5279 | 0.5319 | 0.5359 |
| 0.1 | 0.5398 | 0.5438 | 0.5478 | 0.5517 | 0.5557 | 0.5596 | 0.5636 | 0.5675 | 0.5714 | 0.5753 |
| 0.2 | 0.5793 | 0.5832 | 0.5871 | 0.5910 | 0.5948 | 0.5987 | 0.6026 | 0.6064 | 0.6103 | 0.6141 |
| 0.3 | 0.6179 | 0.6217 | 0.6255 | 0.6293 | 0.6331 | 0.6368 | 0.6406 | 0.6443 | 0.6480 | 0.6517 |
| 0.4 | 0.6554 | 0.6591 | 0.6628 | 0.6664 | 0.6700 | 0.6736 | 0.6772 | 0.6808 | 0.6844 | 0.6879 |
| 0.5 | 0.6915 | 0.6950 | 0.6985 | 0.7019 | 0.7054 | 0.7088 | 0.7123 | 0.7157 | 0.7190 | 0.7224 |
| 0.6 | 0.7257 | 0.7291 | 0.7324 | 0.7357 | 0.7389 | 0.7422 | 0.7454 | 0.7486 | 0.7517 | 0.7549 |
| 0.7 | 0.7580 | 0.7611 | 0.7642 | 0.7673 | 0.7703 | 0.7734 | 0.7764 | 0.7794 | 0.7823 | 0.7852 |
| 0.8 | 0.7881 | 0.7910 | 0.7939 | 0.7967 | 0.7995 | 0.8023 | 0.8051 | 0.8078 | 0.8106 | 0.8133 |
| 0.9 | 0.8159 | 0.8186 | 0.8212 | 0.8238 | 0.8264 | 0.8289 | 0.8315 | 0.8340 | 0.8365 | 0.8389 |
| 1.0 | 0.8413 | 0.8438 | 0.8461 | 0.8485 | 0.8508 | 0.8531 | 0.8554 | 0.8577 | 0.8599 | 0.8621 |
| 1.1 | 0.8643 | 0.8665 | 0.8686 | 0.8708 | 0.8729 | 0.8749 | 0.8770 | 0.8790 | 0.8810 | 0.8830 |
| 1.2 | 0.8849 | 0.8869 | 0.8888 | 0.8907 | 0.8925 | 0.8944 | 0.8962 | 0.8980 | 0.8997 | 0.9015 |
| 1.3 | 0.9032 | 0.9049 | 0.9066 | 0.9082 | 0.9099 | 0.9115 | 0.9131 | 0.9147 | 0.9162 | 0.9177 |
| 1.4 | 0.9192 | 0.9207 | 0.9222 | 0.9236 | 0.9251 | 0.9265 | 0.9279 | 0.9292 | 0.9306 | 0.9319 |
| 1.5 | 0.9332 | 0.9345 | 0.9357 | 0.9370 | 0.9382 | 0.9394 | 0.9406 | 0.9418 | 0.9429 | 0.9441 |
| 1.6 | 0.9452 | 0.9463 | 0.9474 | 0.9484 | 0.9495 | 0.9505 | 0.9515 | 0.9525 | 0.9535 | 0.9545 |
| 1.7 | 0.9554 | 0.9564 | 0.9573 | 0.9582 | 0.9591 | 0.9599 | 0.9608 | 0.9616 | 0.9625 | 0.9633 |
| 1.8 | 0.9641 | 0.9649 | 0.9556 | 0.9664 | 0.9671 | 0.9678 | 0.9686 | 0.9693 | 0.9699 | 0.9706 |
| 1.9 | 0.9713 | 0.9719 | 0.9726 | 0.9732 | 0.9738 | 0.9744 | 0.9750 | 0.9756 | 0.9761 | 0.9767 |
| 2.0 | 0.9772 | 0.9778 | 0.9783 | 0.9788 | 0.9793 | 0.9798 | 0.9803 | 0.9808 | 0.9812 | 0.9817 |
| 2.1 | 0.9821 | 0.9826 | 0.9830 | 0.9834 | 0.9838 | 0.9842 | 0.9846 | 0.9850 | 0.9854 | 0.9857 |
| 2.2 | 0.9861 | 0.9864 | 0.9868 | 0.9871 | 0.9875 | 0.9878 | 0.9881 | 0.9884 | 0.9887 | 0.9890 |
| 2.3 | 0.9893 | 0.9896 | 0.9898 | 0.9901 | 0.9904 | 0.9906 | 0.9909 | 0.9911 | 0.9913 | 0.9916 |
| 2.4 | 0.9918 | 0.9920 | 0.9922 | 0.9925 | 0.9927 | 0.9929 | 0.9931 | 0.9932 | 0.9934 | 0.9936 |
| 2.5 | 0.9938 | 0.9940 | 0.9941 | 0.9943 | 0.9945 | 0.9946 | 0.9948 | 0.9949 | 0.9951 | 0.9952 |
| 2.6 | 0.9953 | 0.9955 | 0.9956 | 0.9957 | 0.9959 | 0.9960 | 0.9961 | 0.9962 | 0.9963 | 0.9964 |
| 2.7 | 0.9965 | 0.9966 | 0.9967 | 0.9968 | 0.9969 | 0.9970 | 0.9971 | 0.9972 | 0.9973 | 0.9974 |
| 2.8 | 0.9974 | 0.9975 | 0.9976 | 0.9977 | 0.9977 | 0.9978 | 0.9979 | 0.9979 | 0.9980 | 0.9981 |
| 2.9 | 0.9981 | 0.9982 | 0.9982 | 0.9983 | 0.9984 | 0.9984 | 0.9985 | 0.9985 | 0.9986 | 0.9986 |
| 3.0 | 0.9987 | | | | | | | | | |

For negative values of $x$, $\Phi(x) = 1 - \Phi(-x)$. For example, let $x = -0.53$. Then $\Phi(-0.53) = 1 - \Phi(0.53) = 1 - 0.7019 = 0.2981$.

# Appendix B: Quantiles of the Chi-Square Distribution

Let $V \sim \chi^2(\nu)$. Then $P\big(V \le q_\beta(\nu)\big) = \beta$.

| $\nu$ | $q_{0.01}(\nu)$ | $q_{0.025}(\nu)$ | $q_{0.05}(\nu)$ | $q_{0.95}(\nu)$ | $q_{0.975}(\nu)$ | $q_{0.99}(\nu)$ |
|---|---|---|---|---|---|---|
| 1 | 0.000157 | 0.000982 | 0.00393 | 3.841 | 5.024 | 6.635 |
| 2 | 0.0201 | 0.0506 | 0.103 | 5.991 | 7.378 | 9.210 |
| 3 | 0.115 | 0.216 | 0.352 | 7.815 | 9.348 | 11.345 |
| 4 | 0.297 | 0.484 | 0.711 | 9.488 | 11.143 | 13.277 |
| 5 | 0.554 | 0.831 | 1.145 | 11.070 | 12.833 | 15.086 |
| 6 | 0.872 | 1.237 | 1.635 | 12.592 | 14.449 | 16.812 |
| 7 | 1.239 | 1.690 | 2.167 | 14.067 | 16.013 | 18.475 |
| 8 | 1.646 | 2.180 | 2.733 | 15.507 | 17.535 | 20.090 |
| 9 | 2.088 | 2.700 | 3.325 | 16.919 | 19.023 | 21.666 |
| 10 | 2.558 | 3.247 | 3.940 | 18.307 | 20.483 | 23.209 |
| 11 | 3.053 | 3.816 | 4.575 | 19.675 | 21.920 | 24.725 |
| 12 | 3.571 | 4.404 | 5.226 | 21.026 | 23.337 | 26.217 |
| 13 | 4.107 | 5.009 | 5.892 | 22.362 | 24.736 | 27.688 |
| 14 | 4.660 | 5.629 | 6.571 | 23.685 | 26.119 | 29.141 |
| 15 | 5.229 | 5.262 | 7.261 | 24.996 | 27.488 | 30.578 |
| 16 | 5.812 | 6.908 | 7.962 | 26.296 | 28.845 | 32.000 |
| 17 | 6.408 | 7.564 | 8.672 | 27.587 | 30.191 | 33.409 |
| 18 | 7.015 | 8.231 | 9.390 | 28.869 | 31.526 | 34.805 |
| 19 | 7.633 | 8.907 | 10.117 | 30.144 | 32.852 | 36.191 |
| 20 | 8.260 | 9.591 | 10.851 | 31.410 | 34.170 | 37.566 |
| 21 | 8.897 | 10.283 | 11.591 | 32.671 | 35.479 | 38.932 |
| 22 | 9.542 | 10.982 | 12.338 | 33.924 | 36.781 | 40.289 |
| 23 | 10.196 | 11.689 | 13.091 | 35.172 | 38.076 | 41.638 |
| 24 | 10.856 | 12.401 | 13.848 | 36.415 | 39.364 | 42.980 |
| 25 | 11.524 | 13.120 | 14.611 | 37.652 | 40.646 | 44.314 |
| 26 | 12.198 | 13.844 | 15.379 | 38.885 | 41.923 | 45.642 |
| 27 | 12.879 | 14.573 | 16.151 | 40.113 | 43.195 | 46.963 |
| 28 | 13.565 | 15.308 | 16.928 | 41.337 | 44.461 | 48.278 |
| 29 | 14.256 | 16.047 | 17.708 | 42.557 | 45.772 | 49.588 |
| 30 | 14.953 | 16.791 | 18.493 | 43.773 | 46.979 | 50.892 |

**Appendix C.** Critical Values $\mathcal{F}_{\nu_1,\nu_2}(\alpha)$ of the $F$ - Distribution

| $\nu_2$ | $\alpha$ | $\nu_1$ | | | | | | | | | | | |
|---|---|---|---|---|---|---|---|---|---|---|---|---|---|
| | | 1 | 2 | 3 | 4 | 5 | 6 | 7 | 8 | 9 | 10 | 15 | 20 |
| 1 | 0.050 | 161.4 | 199.5 | 215.7 | 224.6 | 230.2 | 234.0 | 236.8 | 238.9 | 240.5 | 241.9 | 245.9 | 248.0 |
| 1 | 0.025 | 647.8 | 799.5 | 864.2 | 899.6 | 921.8 | 937.1 | 948.2 | 956.7 | 963.3 | 968.6 | 984.9 | 997.2 |
| 1 | 0.010 | 4052 | 4999 | 5403 | 5625 | 5764 | 5859 | 5928 | 5981 | 6022 | 6056 | 6157 | 6209 |
| 1 | 0.005 | 16211 | 19999 | 21615 | 22500 | 23056 | 23437 | 23715 | 23925 | 24091 | 24224 | 24630 | 24836 |
| 2 | 0.050 | 18.51 | 19.00 | 19.16 | 19.25 | 19.30 | 19.33 | 19.35 | 19.37 | 19.38 | 19.40 | 19.43 | 19.45 |
| 2 | 0.025 | 38.51 | 39.00 | 39.17 | 39.25 | 39.30 | 39.33 | 39.36 | 39.37 | 39.39 | 39.40 | 39.43 | 39.45 |
| 2 | 0.010 | 98.50 | 99.00 | 99.17 | 99.25 | 99.30 | 99.33 | 99.36 | 99.37 | 99.39 | 99.40 | 99.43 | 99.45 |
| 2 | 0.005 | 198.5 | 199.0 | 199.2 | 199.2 | 199.3 | 199.3 | 199.4 | 199.4 | 199.4 | 199.4 | 199.4 | 199.4 |
| 3 | 0.050 | 10.13 | 9.55 | 9.28 | 9.12 | 9.01 | 8.94 | 8.89 | 8.85 | 8.81 | 8.79 | 8.70 | 8.66 |
| 3 | 0.025 | 17.44 | 16.04 | 15.44 | 15.10 | 14.88 | 14.73 | 14.62 | 14.54 | 14.47 | 14.42 | 14.25 | 14.17 |
| 3 | 0.010 | 34.12 | 30.82 | 29.46 | 28.71 | 28.24 | 27.91 | 27.67 | 27.49 | 27.35 | 27.23 | 26.87 | 26.69 |
| 3 | 0.005 | 55.55 | 49.80 | 47.47 | 46.19 | 45.39 | 44.84 | 44.43 | 44.13 | 43.88 | 43.69 | 43.08 | 42.78 |
| 4 | 0.050 | 7.71 | 6.94 | 6.59 | 6.39 | 6.26 | 6.16 | 6.09 | 6.04 | 6.00 | 5.96 | 5.86 | 5.80 |
| 4 | 0.025 | 12.22 | 10.65 | 9.98 | 9.60 | 9.36 | 9.20 | 9.07 | 8.98 | 8.90 | 8.84 | 8.66 | 8.56 |
| 4 | 0.010 | 21.20 | 18.00 | 16.69 | 15.98 | 15.52 | 15.21 | 14.98 | 14.80 | 14.66 | 14.55 | 14.20 | 14.02 |
| 4 | 0.005 | 31.33 | 26.28 | 24.26 | 23.15 | 22.46 | 21.97 | 21.62 | 21.35 | 21.14 | 20.97 | 20.44 | 20.17 |
| 5 | 0.050 | 6.61 | 5.79 | 5.41 | 5.19 | 5.05 | 4.95 | 4.88 | 4.82 | 4.77 | 4.74 | 4.62 | 4.56 |
| 5 | 0.025 | 10.01 | 8.43 | 7.76 | 7.39 | 7.15 | 6.98 | 6.85 | 6.76 | 6.68 | 6.62 | 6.43 | 6.33 |
| 5 | 0.010 | 16.26 | 13.27 | 12.06 | 11.39 | 10.97 | 10.67 | 10.46 | 10.29 | 10.16 | 10.05 | 9.72 | 9.55 |
| 5 | 0.005 | 22.78 | 18.31 | 16.53 | 15.56 | 14.94 | 14.51 | 14.20 | 13.96 | 13.77 | 13.62 | 13.15 | 12.90 |
| 6 | 0.050 | 5.99 | 5.14 | 4.76 | 4.53 | 4.39 | 4.28 | 4.21 | 4.15 | 4.10 | 4.06 | 3.94 | 3.87 |
| 6 | 0.025 | 8.81 | 7.26 | 6.60 | 6.23 | 5.99 | 5.82 | 5.70 | 5.60 | 5.52 | 5.46 | 5.27 | 5.17 |
| 6 | 0.010 | 13.75 | 10.92 | 9.78 | 9.15 | 8.75 | 8.47 | 8.26 | 8.10 | 7.98 | 7.87 | 7.56 | 7.40 |
| 6 | 0.005 | 18.63 | 14.54 | 12.92 | 12.03 | 11.46 | 11.07 | 10.79 | 10.57 | 10.39 | 10.25 | 9.81 | 9.59 |
| 7 | 0.050 | 5.59 | 4.74 | 4.35 | 4.12 | 3.97 | 3.87 | 3.79 | 3.73 | 3.68 | 3.64 | 3.51 | 3.44 |
| 7 | 0.025 | 8.07 | 6.54 | 5.89 | 5.52 | 5.29 | 5.12 | 4.99 | 4.90 | 4.82 | 4.76 | 4.57 | 4.47 |
| 7 | 0.010 | 12.25 | 9.55 | 8.45 | 7.85 | 7.46 | 7.19 | 6.99 | 6.84 | 6.72 | 6.62 | 6.31 | 6.16 |
| 7 | 0.005 | 16.24 | 12.40 | 10.88 | 10.05 | 9.52 | 9.16 | 8.89 | 8.68 | 8.51 | 8.38 | 7.97 | 7.75 |

142

$\nu_1$

| $\nu_2$ | $\alpha$ | 1 | 2 | 3 | 4 | 5 | 6 | 7 | 8 | 9 | 10 | 15 | 20 |
|---|---|---|---|---|---|---|---|---|---|---|---|---|---|
| 8 | 0.050 | 5.32 | 4.46 | 4.07 | 3.84 | 3.69 | 3.58 | 3.50 | 3.44 | 3.39 | 3.35 | 3.22 | 3.15 |
| 8 | 0.025 | 7.57 | 6.06 | 5.42 | 5.05 | 4.82 | 4.65 | 4.53 | 4.43 | 4.36 | 4.30 | 4.10 | 4.00 |
| 8 | 0.010 | 11.26 | 8.65 | 7.59 | 7.01 | 6.63 | 6.37 | 6.18 | 6.03 | 5.91 | 5.81 | 5.52 | 5.36 |
| 8 | 0.005 | 14.69 | 11.04 | 9.60 | 8.81 | 8.30 | 7.95 | 7.69 | 7.50 | 7.34 | 7.21 | 6.81 | 6.61 |
| 9 | 0.050 | 5.12 | 4.26 | 3.86 | 3.63 | 3.48 | 3.37 | 3.29 | 3.23 | 3.18 | 3.14 | 3.01 | 2.94 |
| 9 | 0.025 | 7.21 | 5.71 | 5.08 | 4.72 | 4.48 | 4.32 | 4.20 | 4.10 | 4.03 | 3.96 | 3.77 | 3.67 |
| 9 | 0.005 | 10.56 | 8.02 | 6.99 | 6.42 | 6.06 | 5.80 | 5.61 | 5.47 | 5.35 | 5.26 | 4.96 | 4.81 |
| 9 | 0.001 | 13.61 | 10.11 | 8.72 | 7.96 | 7.47 | 7.13 | 6.88 | 6.69 | 6.54 | 6.42 | 6.03 | 5.83 |
| 10 | 0.050 | 4.96 | 4.10 | 3.71 | 3.48 | 3.33 | 3.22 | 3.14 | 3.07 | 3.02 | 2.98 | 2.85 | 2.77 |
| 10 | 0.025 | 6.94 | 5.46 | 4.83 | 4.47 | 4.24 | 4.07 | 3.95 | 3.85 | 3.78 | 3.72 | 3.52 | 3.42 |
| 10 | 0.010 | 10.04 | 7.56 | 6.55 | 5.99 | 5.64 | 5.39 | 5.20 | 5.06 | 4.94 | 4.85 | 4.56 | 4.41 |
| 10 | 0.005 | 12.83 | 9.43 | 8.08 | 7.34 | 6.87 | 6.54 | 6.30 | 6.12 | 5.97 | 5.85 | 5.47 | 5.27 |
| 11 | 0.050 | 4.84 | 3.98 | 3.59 | 3.36 | 3.20 | 3.09 | 3.01 | 2.95 | 2.90 | 2.85 | 2.72 | 2.65 |
| 11 | 0.025 | 6.72 | 5.26 | 4.63 | 4.28 | 4.04 | 3.88 | 3.76 | 3.66 | 3.59 | 3.53 | 3.33 | 3.23 |
| 11 | 0.010 | 9.65 | 7.21 | 6.22 | 5.67 | 5.32 | 5.07 | 4.89 | 4.74 | 4.63 | 4.54 | 4.25 | 4.10 |
| 11 | 0.005 | 12.23 | 8.91 | 7.60 | 6.88 | 6.42 | 6.10 | 5.86 | 5.68 | 5.54 | 5.42 | 5.05 | 4.86 |
| 12 | 0.050 | 4.75 | 3.89 | 3.49 | 3.26 | 3.11 | 3.00 | 2.91 | 2.85 | 2.80 | 2.75 | 2.62 | 2.54 |
| 12 | 0.025 | 6.55 | 5.10 | 4.47 | 4.12 | 3.89 | 3.73 | 3.61 | 3.51 | 3.44 | 3.37 | 3.18 | 3.07 |
| 12 | 0.010 | 9.33 | 6.93 | 5.95 | 5.41 | 5.06 | 4.82 | 4.64 | 4.50 | 4.39 | 4.30 | 4.01 | 3.86 |
| 12 | 0.005 | 11.75 | 8.51 | 7.23 | 6.52 | 6.07 | 5.76 | 5.52 | 5.35 | 5.20 | 5.09 | 4.72 | 4.53 |
| 13 | 0.050 | 4.67 | 3.81 | 3.41 | 3.18 | 3.03 | 2.92 | 2.83 | 2.77 | 2.71 | 2.67 | 2.53 | 2.46 |
| 13 | 0.025 | 6.41 | 4.97 | 4.35 | 4.00 | 3.77 | 3.60 | 3.48 | 3.39 | 3.31 | 3.25 | 3.05 | 2.95 |
| 13 | 0.010 | 9.07 | 6.70 | 5.74 | 5.21 | 4.86 | 4.62 | 4.44 | 4.30 | 4.19 | 4.10 | 3.82 | 3.66 |
| 13 | 0.005 | 11.37 | 8.19 | 6.93 | 6.23 | 5.79 | 5.48 | 5.25 | 5.08 | 4.94 | 4.82 | 4.46 | 4.27 |
| 14 | 0.050 | 4.60 | 3.74 | 3.34 | 3.11 | 2.96 | 2.85 | 2.76 | 2.70 | 2.65 | 2.60 | 2.46 | 2.39 |
| 14 | 0.025 | 6.30 | 4.86 | 4.24 | 3.89 | 3.66 | 3.50 | 3.38 | 3.29 | 3.21 | 3.15 | 2.95 | 2.84 |
| 14 | 0.010 | 8.86 | 6.51 | 5.56 | 5.04 | 4.69 | 4.46 | 4.28 | 4.14 | 4.03 | 3.94 | 3.66 | 3.51 |
| 14 | 0.005 | 11.06 | 7.92 | 6.68 | 6.00 | 5.56 | 5.26 | 5.03 | 4.86 | 4.72 | 4.60 | 4.25 | 4.06 |
| 15 | 0.050 | 4.54 | 3.68 | 3.29 | 3.06 | 2.90 | 2.79 | 2.71 | 2.64 | 2.59 | 2.54 | 2.40 | 2.33 |
| 15 | 0.025 | 6.20 | 4.77 | 4.15 | 3.80 | 3.58 | 3.41 | 3.29 | 3.20 | 3.12 | 3.06 | 2.86 | 2.76 |
| 15 | 0.010 | 8.68 | 6.36 | 5.42 | 4.89 | 4.56 | 4.32 | 4.14 | 4.00 | 3.89 | 3.80 | 3.52 | 3.37 |
| 15 | 0.005 | 10.80 | 7.70 | 6.48 | 5.80 | 5.37 | 5.07 | 4.85 | 4.67 | 4.54 | 4.42 | 4.02 | 3.88 |

# References

Analytical Software. 2000. *Statistix 7, Users Manual.* Analytical Software, Tallahassee, FL.

Beauregard, M., R. Mikulak and B. Olson. 1992. *A Practical Guide to Quality Improvement.* Van Nostrand Reinhold.

Bisgaard, S. 2002. An investigation involving the comparison of several objects. *Quality Engineering*, **14**(4), 685–689.

Bolshev, L.N. and N.V. Smirnov. 1965. *Tables of Mathematical Statistics.* Nauka (in Russian).

Box. G. 1953. Nonnormality and tests on variances. *Biometrika*, **40**, 318–335.

Box, G. 1998. Multiple sources of variation: variance components. *Quality Engineering*, **11**(1), 171–174.

Box, G. and A. Luceno. 1997. *Statistical Control by Monitoring and Feedback Adjustment.* John Wiley & Sons, Inc.

Cochran, W.G. and G.M. Cox. 1957. *Experimental Designs*, 2nd ed. John Wiley & Sons, Inc.

Cramér, H. 1946. *Mathematical Methods of Statistics.* Princeton University Press, Princeton.

Veronica Czitrom and Patrick D. Spagon. 1997. *Statistical Case Studies for Industrial Process Improvement.* SIAM – ASA.

Dechert, J. et al. 2000. Statistical process control in the presence of large measurement variation. *Quality Engineering*, **12**(3), 417–423.

DeGroot, M.H. 1975. *Probability and Statistics*, 2nd ed. Addison-Wesley.

Devore, J.L. 1982. *Probability and Statistics for Engineering and Sciences.* Brooks/Cole Publishing Company, Monterey California.

Dunin-Barkovsky, I.V. and N.V. Smirnov. 1955. *Theory of Probability and Mathematical Statistics for Engineering.* Gosizdat, Moscow (in Russian).

Eurachem. 2000. *Quantifying Uncertainty in Chemical Measurement.* Eurachem Publications.

Fisher, R.A. 1941. *Statistical Methods for Research Workers.* G.E. Stechert & Co., New York.

Andrew Gelman, John B. Carlin, Hal S. Stern and Donald Rubin. 1995. *Bayesian Data Analysis.* Chapman & Hall/ CRC.

Gertsbakh, I.B. 1989. *Statistical Reliability Theory.* Marcel Dekker, Inc.

Grubbs, F.E. 1948. On estimating precision of measuring instruments and product variability. *Journal of the American Statistical Association,* **43**, 243–264.

Grubbs, F.E. 1973. Comparison of measuring instruments. *Technometrics,* **15**, 53–66.

Hald, A. 1952. *Statistical Theory With Engineering Applications.* Wiley, New-York – London.

Hurwitz et al. (1997) Identifying the sources of variation in wafer planarization process, in V. Czitrom and P.D. Spagon (eds), *Statistical Case Studies for Industrial Process Improvement,* pp. 106–113. Society for Industrial and Applied Mathematics and American Statistical Association.

Kotz, S. and N.L. Johnson (eds). 1983. *Encyclopedia of Statistical Sciences,* Vol. 3, pp. 542–549. Wiley.

Kragten, J. 1994. Calculating standard deviations and confidence intervals with a universally applicable spreadsheet technique. *Analyst,* **119**, 2161–2166.

Krylov, A.N. 1950. *Lectures on Approximate Calculations.* Gosizdat, Moscow and Leningrad (in Russian).

Madansky, A, 1988. *Prescriptions for Working Statisticians.* Springer, New York.

Mandel, J. 1991. *Evaluation and Control of Measurements.* Marcel Dekker, Inc.

Miller, J.C. and J.N. Miller. 1993. *Statistics for Analytical Chemistry,* 3rd ed. Ellis Hoorwood PTR Prentice Hall, New York.

Mitchell,T, Hegeman, V. and K.C. Liu. 1997. GRR methodology for destructive testing, in V. Czitrom and P.D. Spagon (eds), *Statistical Case Studies for Industrial Process Improvement,* pp. 47–59. Society for Industrial and Applied

Mathematics and American Statistical Association.

Montgomery, D.C. 2001. *Design and Analysis of Experiments*, 5th ed. John Wiley & Sons, Inc.

Montgomery, D.C. and G.C. Runger. 1993. Gauge capability and designed experiments. Part I: basic methods. *Quality Engineering*, **6**(1), 115–135.

Morris, A.S. 1997. *Measurement and Calibration Requirements for Quality Assurance ISO 9000*. John Wiley & Sons, Inc.

Nelson, L.S. 1984. The Shewhart control chart – tests for special causes. *Journal of Quality Technology*, **16**, 237–239.

*NIST Technical Note 1297*. 1994. Guidelines for Evaluating and Expressing the Uncertainty of NIST Measurement Results.

Pankratz, P.C. 1997. Calibration of an FTIR spectrometer for measuring carbon, in V. Czitrom and P.D. Spagon (eds), *Statistical Case Studies for Industrial Process Improvement*, pp. 19–37. Society for Industrial and Applied Mathematics and American Statistical Association.

Rabinovich, S. 2000. *Measurement Errors and Uncertainties: Theory and Practice*, 2nd ed. Springer, New York.

Sachs, L. 1972. *Statistische Auswertungsmethoden*, 3rd ed. Springer, Berlin Heidelberg New York.

Sahai, H. and M.I. Ageel. 2000. *The Analysis of Variance*. Birkhäuser, Boston Basel Berlin.

Scheffé, H. 1960. *The Analysis of Variance*. John Wiley & Sons, Inc. New York.

Stigler, S.M. 1977. Do robust estimators work with real data? (with discussion). *Annals of Statistics*, **45**, 179–183.

Taylor, J.R. 1997. *An Introduction to Error Analysis*. University Science Books, Susalito, California.

Vardeman, S.B. and E.S. VanValkenburg. 1999. Two-way random-effects analyses and gauge R&R studies. *Technometrics*, **41**(3), 202–211.

Wolfram, Stephen. 1999. *The Mathematica Book*, 4th ed. Cambridge University Press.

Youden, W.J. and E.H. Steiner. 1975. *Statistical Manual of the Association of Official Analytic Chemists: Statistical Techniques for Collaborative Studies*. AOAC International, Arlington, VA.

# Index